Mir Mohammad Alavi Nikje
Recycling of Polyurethane Wastes

Also of interest

Recycling of Polyethyleneterephthalate
Forrest, 2019
ISBN 978-3-11-064029-8, e-ISBN 978-3-11-064030-4

Polymer Engineering
Tylkowski, Wieszczycka, Jastrzab (Eds.), 2017
ISBN 978-3-11-046828-1, e-ISBN 978-3-11-046974-5

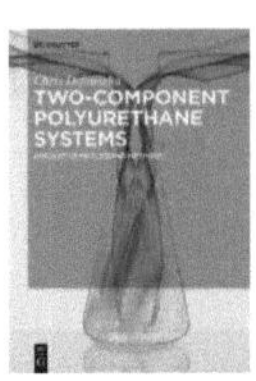

Two-Component Polyurethane Systems.
Innovative Processing Methods
Defonseka, 2019
ISBN 978-3-11-063957-5, e-ISBN 978-3-11-064316-9

Flexible Polyurethane Foams.
A Practical Guide
Defonseka, 2019
ISBN 978-3-11-063958-2, e-ISBN 978-3-11-064318-3

e-Polymers.
Editor-in-Chief: Seema Agarwal
ISSN 2197-4586
e-ISSN 1618-7229

Mir Mohammad Alavi Nikje

Recycling of Polyurethane Wastes

2nd Edition

DE GRUYTER

Author
Prof. Dr. Mir Mohammad Alavi Nikje
Department of Chemistry
Faculty of Science
Imam Khomeini International University
drmm.alavi@gmail.com

ISBN 978-3-11-064102-8
e-ISBN (PDF) 978-3-11-064159-2
e-ISBN (EPUB) 978-3-11-064126-4

Library of Congress Control Number: 2018966617

Bibliographic information published by the Deutsche Nationalbibliothek
The Deutsche Nationalbibliothek lists this publication in the Deutsche Nationalbibliografie; detailed bibliographic data are available on the Internet at http://dnb.dnb.de.

Typesetting: Integra Software Services Pvt. Ltd.
Printing and binding: CPI books GmbH, Leck
Cover image: onurdongel/getty images

www.degruyter.com

Preface

Polyurethanes (PU) represent a very broad family of polymers, from flexible/rigid foams, thermoplastic elastomers to adhesives, paints and varnishes.

Treatments of wastes produced during processes as well as aged polymeric materials have been a problem worldwide, and trends suggest that the overall amount of solid waste that humans generate will continue to increase. PU are used in many everyday applications and have industrial uses. Hence, they enter municipal solid-waste streams, usually by discarding of consumer and industrial products. Frequently, these products are durable goods with a long lifespan (e.g., upholstered furniture, mattresses and automobile parts).

PU recycling is done to minimise waste and reduce environmental pollution. The other major advantage of recycling for waste processing is that of economy and investment cost. Because of the lack of economic efforts for recyclate marketing, research into PU recycling is increasing. Unfortunately, and specifically in undeveloped and third-world countries, industries prefer landfilling of waste as a method of recycling!

Recycling routes for PU can be divided into mechanical and chemical methods. In the latter, glycolysis is one of the most important methods that is used on a commercial scale. The reaction products in glycolysis are polyols, and used directly for the manufacture of new PU. The recovered polyol may replace ≤50% of virgin polyol material for formulation of PU foams. By recycling and recovering polyols from waste PU, new materials can be produced. This process can reduce environmental pollution and aid job creation.

In this book, these methods are investigated to find a suitable method for waste reduction, protection of the environment and prevention of waste landfilling. PU recycling has posed unique challenges due to its wide variety of applications. Recycling has opened an effective and economic route for treatment of PU waste. Nonetheless, more research efforts are required to scale-up recycling methods.

The book reviews aspects from contemporary literature (including our research) focused on these topics. Recent progress has been done by the author's research team in chemical recycling of PU waste. Recycling methods are stated clearly, as is use of instrumental methods such as nuclear magnetic resonance spectroscopy and Fourier-Transform infrared spectroscopy for characterisation and identification of recycling products. This book provides new and exciting vistas for finding adequate recycling methods as well as the starting materials and intermediates for PU products.

Mir Mohammad Alavi Nikje,
Department of Chemistry, Faculty of Science, Imam Khomeini International University, P.O. Box 34148 – 96818, Qazvin, Iran

https://doi.org/10.1515/9783110641592-201

Acknowledgements

I would like to express my gratitude to the many people who helped me while I wrote this book.

I would like to proffer my deepest thanks to my University, the Imam Khomeini International University in Qazvin, Iran, which provided financial support for my research projects.

I must also thank all of my students and artwork reviewers who reviewed our articles, offered comments, and allowed me to quote their remarks and assisted in improvements of my studies.

Above all, I want to thank my wife, and my two children (Arian and Yekta), who supported and encouraged me even though it meant spending less time with them.

Mir Mohammad Alavi Nikje

https://doi.org/10.1515/9783110641592-202

Contents

1 Introduction

1.1 Introduction to polyurethanes

The German chemist Otto Bayer invented polyurethane (PU) in the 1930s as part of his research on polymer fibres. PU is produced by reacting a polyether or polyester polyol as a hydroxyl-containing monomer (a polymeric alcohol with more than two reactive hydroxyl groups per molecule), such as polypropylene glycol or polytetramethylene glycol, with diisocyanate or a polymeric isocyanate [e.g., 4,4′-diphenylmethane diisocyanate (MDI) or toluene diisocyanate (TDI)] in the presence of suitable catalysts and additives, especially chain extenders (e.g., 1,4-butanediol (1,4-BD) [1, 2]) as shown in Equation 1.1:

$$R - NCO + OH - R \rightarrow R - NH - COOR \tag{1.1}$$

With a wide range of high-performance properties, PU are essential for a multitude of end-use applications. PU represent an important class of polymers that have found widespread use in medical, automotive and industrial fields, and are one of the most important classes of polymeric materials within the plastics 'family'. Due to the wide versatility of the materials, PU is around us in many products we use every day. For example, it can be found in liquid form in coatings and paints, and in the solid state as elastomers, rigid insulation for buildings, soft flexible foam in mattresses and automotive seats, or as an integral 'skin' in sports goods such as skis, surfboards and automotive parts. Obviously, because of the large variety of starting materials (i.e., polyols and isocyanates), this family would exhibit various organic chemical behaviours. By considering endless chemical-processing routes of the reacting chemicals, numerous various materials may be synthesised that would fall under the definition of PU. Besides urethane linkages, PU structure may be defined by ether, urea, and amide functional groups. As a result, PU stands for a product range or plastics-industry segment rather than for a single, well-defined polymer resin (e.g., polyvinyl chloride). The range of applications of the various PU materials is extensive. A simple classification of PU-based products is given in Table 1.1.

As shown in the Table 1.1, rigid and flexible PU foams (PUF) are produced in large quantities and consumed in various industries. The ratio of application of rigid to flexible foam is ≈1:3. PUF are defined depending on whether the cell structure is 'open' or 'closed'.

In rigid PUF, a low percentage of cells have an open-cell structure, and the bulk density of these products is ≈30–35 kg/m^3. The physical blowing agent as the gas contained within the cells results in very low thermal conductivity, and the main use of rigid PUF is in the insulating panels of refrigerators. Flexible PUF have a virtually completely open-cell structure with typical densities of 20–45 kg/m^3.

https://doi.org/10.1515/9783110641592-001

Table 1.1: Classification of PU-based products according to density and rigidity.

Density	Rigidity		
	High	**Medium**	**Low**
High	Glass fibre RIM-PUR	Elastomer, TPU, RIM-PU	–
Medium	Rigid integral skin foam	–	Flexible integral skin foam
Low	Flexible foam	Filling foam	Rigid foam

RIM: Reaction injection moulding
TPU: Thermoplastic polyurethane
Reproduced with permission from M.M. Alavi Nikje,
A. Bagheri Garmarudi and A.B Idris, *Designed Monomers and Polymers*, 2011, **14**, 5, 395.

Therefore, they are not useful as insulators and are often used in the production of seats and furniture [1].

During recent decades, chlorofluorocarbons (CFC) such as CFC-11 (CCl_3F) have been used widely as physical blowing agents because they are non-toxic, non-flammable, stable, inexpensive, very soluble in isocyanate and polyols but not in the PU, and give excellent thermal insulation [2]. However, they have ozone-depleting potential and are not environmentally friendly. To overcome these drawbacks, organic liquids with low boiling points have been found to be useful physical blowing agents in foam production, with the additional benefit (discovered when developing rigid PUF) of improving the insulating properties of the final product. PUF production is relatively straightforward. It can be accomplished by adding water as the chemical blowing agent, which results in the release of carbon dioxide as the reaction product of water and isocyanate, and more urea [–(NH–CO–NH)–] linkages in the polymer chain. However, using water alone as the blowing agent would not result in satisfactory quality or durability of the product for many applications.

1.2 Chemistry of polyurethanes

PU are a broad class of polymers that contain the urethane (–NH-CO-O–) linkage or functional groups in the polymer backbone. They are formed by the reaction of an isocyanate containing two or more isocyanate groups per molecule [$R\text{-}(N{=}C{=}O)_n$, $n \geq 2$] with a polyol containing on average of two or more hydroxyl groups per molecule [$R'\text{-}(OH)_n$, $n \geq 2$] in the presence of a catalyst and other additives *via* a well-known gel reaction [2].

1.2.1 Isocyanates

Isocyanates are essential components required for PU production that must have two or more isocyanate groups on each molecule. The most important and common aromatic isocyanates in the PU industry are TDI and MDI, respectively. Industrially, TDI is formed by the phosgenation of diaminotoluene – which is synthesised by the reduction of nitrotoluene – as a mixture of the 2,4- and 2,6-TDI isomers (80/20 TDI). In the same reaction pathway, MDI is formed by the phosgenation of the condensation product of aniline and formaldehyde [3]. In the meantime, the easy availability and low costs of TDI and MDI isocyanates precursors resulted in these isocyanates having the greatest role and consumptions in the PU industry. Industrial grades of TDI and MDI comprising mixtures of isomers often contain polymeric materials. They are used to make flexible foam (e.g., slabstock foam for mattresses or moulded foams for car seats), rigid foam (e.g., insulating foam in refrigerators), and elastomers (e.g., shoe soles). Isocyanates can be modified by partial reactions with polyols or by introduction of other functional groups to reduce volatility and toxicity. These strategies decrease their freezing points to make handling easier or to improve the properties of the final products. The choice of the isocyanate for PU synthesis is governed by the properties required for end-use applications. For the manufacture of rigid PU, aromatic isocyanates are chosen, and PU derived from these isocyanates demonstrate lower oxidative and ultraviolet (UV) stabilities. The other main groups of isocyanates are aliphatic and cycloaliphatic. They are used in limited volumes, most often in coatings and other applications where colour and transparency are important because PU made with aromatic isocyanates tend to darken upon exposure to light. The most important aliphatic and cycloaliphatic isocyanates are 1,6-hexamethylene diisocyanate (HDI), isophorone diisocyanate (IPDI), and 4,4′-diisocyanato dicyclohexylmethane (H_{12}MDI). IPDI is used widely in the preparation of light-stable PU coatings. For a brief review of the chemical structures of common isocyanates, see Figure 1.1.

1.2.2 Polyols

The other component in PU formation is a polyol. The chemical structure of these compounds has a profound effect on the properties of the final PU polymers. These compounds have a hydroxyl group that can react with diisocyanate, which enables preparation of various types of PU. Structurally, polyols are divided into two main categories: polyether polyols (manufactured by the reaction of an epoxy functional group-containing starting material with a reactive hydrogen-containing starter or initiator) and polyester polyols (made by the polycondensation reaction of two and/or multifunctional carboxylic acids and polyhydroxyl compounds or alcohols). Polyols can be classified further according to their end use. Polyester polyols

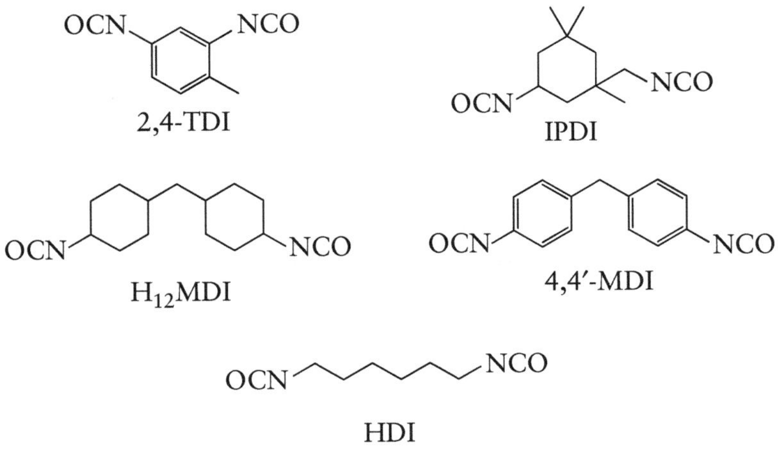

Figure 1.1: Common isocyanates.

consist of ester and hydroxylic groups in one backbone. For example, polycaprolactones (PCL) are polyester polyols produced by ring-opening polymerisation of a caprolactone monomer with a glycol such as diethylene glycol (DEG) or ethylene glycol (EG) and other diols and triols. The wide variety in PU applications refers to the molecular weight (Mw) of the polyol, polyol type, crosslink densities, and isocyanate type. In this case, higher-Mw polyols varying from 2,000 to 10,000 are used for the formulation of more flexible PU, whereas lower-Mw polyols make more rigid products. Monomeric polyols such as glycerin, pentaerythritol (PER), EG and sucrose often serve as the starting materials or initiators for preparation of polymeric polyols.

Flexible PUF comprise polyols with low-functionality (f) initiators such as dipropylene glycol ($f = 2$), glycerine ($f = 3$) or a sorbitol/ water mixture solution ($f = 2.75$). Polyols for rigid applications use high-functionality initiators such as sucrose ($f = 8$), sorbitol ($f = 6$), toluenediamine ($f = 4$), and Mannich bases ($f = 4$). For preparation of all polyether polyols of these categories, propylene oxide (PO)- and/or other epoxy functional group-containing compounds and a mixture of these starting materials are added to initiators until the desired Mw is achieved. The order of addition and the amounts of each epoxy functional group-containing starting material affect the properties of the polyol, such as compatibility, water solubility, and reactivity. For example, polyols prepared with only PO are terminated by secondary hydroxyl groups and are less reactive when compared with the reactivity of polyols capped with ethylene oxide (which contain primary hydroxyl groups). Graft polyols (also called 'filled polyols' or 'polymer polyols') contain finely dispersed styrene- acrylonitrile, acrylonitrile, or polyurea (polyharnstoff) dispersions. Flexible moieties in the final product increase the load-bearing properties of low-density high-resiliency foam, as well as improve the toughness of microcellular foams that are often used as shock absorbers and cast elastomers. Hence, starters or initiators such as ethylenediamine and triethanolamine are used to make low-Mw rigid foam polyols that have catalytic activity due to nitrogen atoms in the backbone that can promote blowing and gel reactions. A special class of polyether polyols called polytetramethylene ether glycols (which are made by polymerising tetrahydrofuran) are used in high-performance coating, wetting and elastomer applications. Figure 1.2 shows examples of some common polyols.

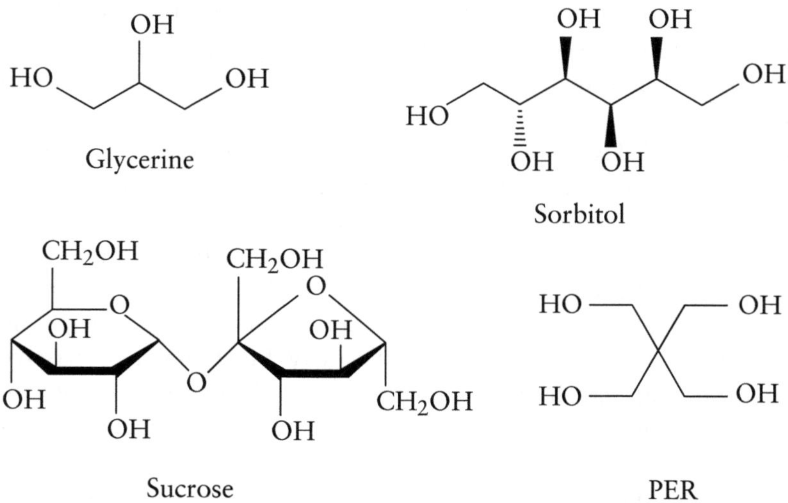

Figure 1.2: Common polyols.

1.2.3 Additives

Catalysts, chain extenders, surfactants, crosslinkers and other additives may be required during the synthesis of PU. In general, tertiary amines and organotin compounds are used as catalysts in the synthesis of PU. Difunctional low-Mw diols such as EG, DEG, and 1,3-propylene glycol 1,4-BD are used as chain extenders in PU synthesis. 1,4-BD is the most important chain extender in the synthesis of PU elastomers. Low-Mw triols (compounds with higher functionality) such as glycerol, trimethylolpropane, tetrol (e.g., PER), diamines (e.g., hexamethylene diamine), and aminoalcohol (e.g., diethanolamine) are considered to be crosslinkers [3].

1.3 Brief studies of polyurethane types

The variety of PU types ranges from flexible/rigid foams and thermoplastic elastomers to adhesives, paints and varnishes.

All types of PUF are formed by the appropriate choice of polyol and isocyanate with respect to chemical structure, equivalent weight, and functionality.

1.3.1 Flexible polyurethane foams

Morphology studies of flexible PUF have shown predominantly open-cells that allow free movement of air or 'breathing' if the foam flexed. Because of this characteristic, flexible foams show excellent sound-absorbing properties and low thermal conductivity. Unfortunately, they are discoloured rapidly by exposure to UV radiation.

The other family of flexible foams (also termed 'semi-flexible') is integral skin foams. Integral skin foam consists of a high-density skin foam and low-density core foam. Due to appropriate wear resistances and wide-ranging physical characteristics, they have found wide application in the automotive industry. New styling trends, surface design, safety functions, and comfort are made possible by these materials. The interior of cars, trucks, and public transportation (bus, rail, and airplane) must satisfy functional and passive safety requirements (e.g., crash energy should be absorbed reversibly by protection padding). Steering wheels manufactured with integral skin foam have found wide application in car manufacturing [4].

Foams of semi-flexible type are prepared using appropriate combinations of polyesters and isocyanates. Semi-flexible foams offer high modulus and good shock-absorbing qualities. These PUF are composed of open-cells. Acoustic insulation is one of the advantages of an open-cell structure.

1.3.2 Rigid polyurethane foams

Rigid PUF constitutes an important group of PUF. Demand is increasing rapidly because of wide applications as a heat insulator in refrigerators, buildings and construction materials, floatation and packaging/padding materials, encapsulates to isolate and support thermally sensitive components within weapon systems, and shock absorbers in industry.

Rigid PUF is a heat-insulating material and is one of the principal components of home appliances or heat-insulated systems such as water pipes. Foam properties are affected by the properties of raw materials and are altered by modifiers. The furniture industry, heat- insulation systems, building construction, transportation (including automotives), the shoe industry, and aviation use PUF. Unlike flexible foams, rigid PUF consist of a high percentage of closed-cells. Rigid PUF has a low density (25–150 kg/m^3), so a small amount of it occupies large spaces that can cause environment pollution [5].

1.3.3 Polyurethane elastomers

PU elastomers are polymers with considerable versatility. Through appropriate choice of the diisocyanates and diols that combine in their synthesis, a wide range of physical properties are achievable.

PU elastomers are linear block copolymers. They contain soft segments based on macrodiol (polyether, polyester or polybutadiene polyols) and hard segments (isocyanates and chain extenders). PU elastomers are generated in a wide variety from cast elastomers to TPU (which are synthesised mainly from polyester diols or PCL [6]).

1.4 Polyurethane market, consumption and environmental impact

PU products are an important part of everyday life and are similar to other petrol-derived polymers. Recently, public awareness of environmental issues has increased enormously: the world has limited natural resources and a limited capacity to manage the volume of waste that human activities generate. As the volume of PU production increases, so does the amount of PU waste.

Due to the variety of physical, chemical, thermal and mechanical properties, PU have a wide range of applications:

- Flexible PUF: automobile seating, furniture, carpets;
- Rigid PUF: refrigerators, insulation boards;
- Elastomers: footwear, adhesives, medical;
- Reaction injection moulding: automobiles (bumpers, side panels); and
- Other: carpets, casting, sealants [7].

This wide variety of applications results in a large amount of consumption and causes environmental problems if PU remain in the ecosystem (Table 1.2). Because of the huge amount of production of polymeric wastes, treatment of aged polymeric materials has become a problem worldwide. PU use in everyday life and industry is unavoidable, so they enter municipal solid-waste streams (usually by discarding of consumer and industrial products). These products often are durable goods with a long lifespan (e.g., upholstered furniture, mattresses, automobile parts).

Literature surveys and reviews of news reveal that the total production of PU in the Euro zone, Middle East, Africa and the American continent are increasing annually. For example, total production of PU in the Europe, Middle East and Africa (EMEA) region was 5,756,350 tonnes in 2013 compared with ≈5.5 million tonnes in 2011 and 5.1 million tonnes in 2009. By 2018, despite continuing concerns surrounding the global economy, this figure is forecast to be 6,630,335 tonnes, equivalent to an average growth of 2.9% annually and over the next 5 years. Total production of flexible and rigid PUF

Table 1.2: Total production of all PU products and growth rates by region, 2013–2018.

Region/Country	2013 (tonnes)	2018 (tonnes)	2013–2018 (% growth p.a.)
Western Europe	3,347,360	3,680,560	1.9
Eastern Europe	1,434,330	1,726,505	3.8
Middle East and Africa	974,660	1,223,270	4.6
North America	3,203,490	3,804,870	3.5
South America	776,280	976,415	4.7

Reproduced with permission from IAL Consultants [8]

in 2013 in the EMEA region was ≈2.00 and ≈1.85 million tonnes, respectively. Hence, total production of rigid PUF in the EMEA increased from 1.62 million tonnes in 2011 to 1.85 million tonnes in 2013, of which the bulk was used for thermal insulation in construction and refrigeration industries in the form of panels, boards and *in situ* foam productions. Beside foams, the extremely high performance abilities of PU elastomers, use of this family of compounds has increased in various sectors (e.g., automotive, construction, mining, aerospace, electronics, medical devices, sports products, and fibres) and production in the EMEA was ≈451,000 tonnes in 2013.

Total production of PU products in the Asia-Pacific region was reported to have been 9.83 million tonnes in 2012 compared with 6.94 million tonnes in 2009. This figure is forecast to be 15.08 million tonnes in in 2017, equivalent to an average growth of 8.9% annually over the next 5 years. Total production of flexible PUF in 2012 in the Asia-Pacific region was estimated to have been 2.1 million tonnes, and has been forecast to be 2.9 million tonnes by 2017. Total production of rigid PUF in the Asia-Pacific region has increased from 1.04 million tonnes in 2009 to an estimated 1.82 million tonnes in 2012, of which the bulk was used for thermal insulation in the refrigeration industry. Industrial Aids Limited (IAL) predicts that production of rigid foam will increase to 2.36 million tonnes by 2017. Total production of PU products in the Americas was reported to have been 3.98 million tonnes in 2013 compared with 3.84 million tonnes in 2012. By 2018, this figure is forecast to be 4.78 million tonnes, equivalent to an average growth of 3.7% annually over the next five years [8]. These data mean that scientists need to find environmentally friendly, sustainable, and benign methods for waste treatments.

References

1. M.M. Alavi Nikje, A. Bagheri Garmarudi and A.B. Idris, *Designed Monomers and Polymers*, 2011, **14**, 5, 395.
2. M. Szycher in *Szycher's Handbook of Polyurethanes*, CRC Press, Boca Raton, FL, USA, 1999.

3. K. Ashida in *Polyurethane and Related Foams: Chemistry and Technology*, CRC Press, Taylor & Francis Group, Boca Raton, FL, USA, 2006.
4. M.M. Alavi Nikje, M. Haghshenas and Bagheri Garmarudi, A. Bagheri Garmarudi *Polymer–Plastics Technology and Engineering*, 2006, **45**, 4, 569.
5. M.M. Alavi Nikje and M. Nikrah, *Journal of Macromolecular Science, Part A: Pure and Applied Chemistry*, 2007, **44**, 6, 613.
6. C. Prisacariu in *Polyurethane Elastomers: From Morphology to Mechanical Aspects*, Springer, New York, NY, USA, 2011.
7. Gobi in *New Forecasts for Polypropylene, Polystyrene and Polyurethane*, Gobi International, London, UK, 2002.
8. http://www.ialconsultants.com.

2 Brief review of the methods of recycling of polyurethane foam wastes

2.1 Recycling methods

As mentioned in Chapter 1, polyurethanes (PU) are one of the most versatile groups of plastic materials, with a variety that stretches from flexible/rigid foams and thermoplastic elastomers to adhesives, paints, and varnishes. This variety is one of the reasons why development of cost-efficient recycling methods is very difficult. PU production is expensive compared with mass-produced plastic materials (e.g., polyolefin plastics), which is the reason for the development of recycling methods (i.e., physical, chemical and thermochemical and energy recovery) for PU wastes since the 1960s.

Recycling routes for PU can be divided into 'mechanical (physical)', 'chemical' and 'energy'. Mechanical recycling covers grinding, compression moulding, adhesive pressing and bonding of PU wastes. Chemical methods (also called 'feedstock recycling') change the chemistry of the materials and refer to the breakdown of targeted bonds to recover valuable materials. Energy recovery refers to incineration of PU waste, decomposition (by pyrolysis or hydrogenation) and the combustion of products.

2.2 Mechanical recycling

Mechanical (physical) recycling of PU wastes refers to one of the many forms of 'particle recycling': regrinding, rebinding, adhesive pressing, injection moulding and compression moulding. Grinding of PU wastes into powders and then reuse of these powders in different ways constitutes a mechanical recycling approach. Usually, the wastes for this process come from factory trims and scraps from the production line, as well as post-consumer products. The resultant powder can be used as filler for generation of PU foams (PUF) or elastomers. In this case, the powders are usually added first to the polyol component in a PU production process. Moulded PU products can contain ≤20% regrind material without serious deterioration in physical properties, mechanical properties or performance. To be used as filler, the particle size of the PU powder should be <100 and 200 μm, and different milling and knife-cutting processes are needed.

In another process, shredded polyurethane foam (PUF) wastes can be rebound using pressure or heat. An adhesive binder is used to make vibration- and sound-dampening mats, floorings, sports mats, cushioning and carpet underlay. In a similar method known as 'adhesive pressing', PU granules in different shapes and sizes are coated with a curable binder and cured under pressure and heat, and are used predominately as automotive floor mats and tire covers. In another physical

https://doi.org/10.1515/9783110641592-002

recycling method of PU wastes, reaction injection moulding (RIM) and reinforced-RIM parts can be ground into small particles, which can be moulded under high pressure and heat to form solid parts for the automotive industry. In addition, compression moulded solid parts such as pump and motor housings can contain ≤100% RIM regrind [1].

2.2.1 Rebinding

In rebinding, the final product is moulded PU from pieces of shredded flexible PUF, held together with a binder. Its relatively high density and excellent resilience make it appropriate for vibration- and sound-dampening, flooring, sport mats, cushioning, packaging and carpet underlays [1, 2]. Rebinding has been used for decades. Database surveys show that ≤50,000 tonnes of rebinding foams received from wastes are processed each year in Western Europe, and new applications are being developed. PUF scrap can be reprocessed by mixing scrap particles of particle size ≈1 cm with an appropriate diisocyanate [e.g., 4,4′-diphenylmethane diisocyanate (MDI)] followed by the form-shaping at 30–200 bar and 100–200 °C. In this process, PU-based construction boards with high resistance to water and moisture as well as insulation panels for use in new freezers or refrigerators can be obtained. In the same process, PUF can also be recycled to blocks that find use in carpets, sports mats or furniture. For example, the enormous amount of PUF recovered from scrapped vehicles may satisfy a great part (≈50% in the USA) of the market for carpet underlay. This recycling option is very important for scrapped PUF received from buildings and constructions but is more complicated (and frequently impossible) due to the use of flame-retardants in these materials. Rebinding of PUF can be carried out through batch moulding or continuous moulding, and the foam blocks are processed further to make parts and articles, resulting in trim which in turn can be reused in this process. One of the most important benefits in the mould–shape rebinding process is optimisation of material use and costs. In conclusion, the batch moulding rebinding process comprises collection → sorting → shredding → coating with binder → compressing shaping → activation of adhesive binder → curing of adhesive binder → fabrication of rebounded parts (Figure 2.1) [1, 3].

The quality of the final product of rebinding is dependent upon: types and grades of the foams used; particle size and uniformity of foam pieces; required density of the final product; quality of the binder as well as binder to foam ratio.

2.2.2 Regrinding

Regrind technology (also termed ‘powdering’) involves reuse of ground PU waste as filler in the PUF or elastomer formulation and involves (i) grinding PU material into

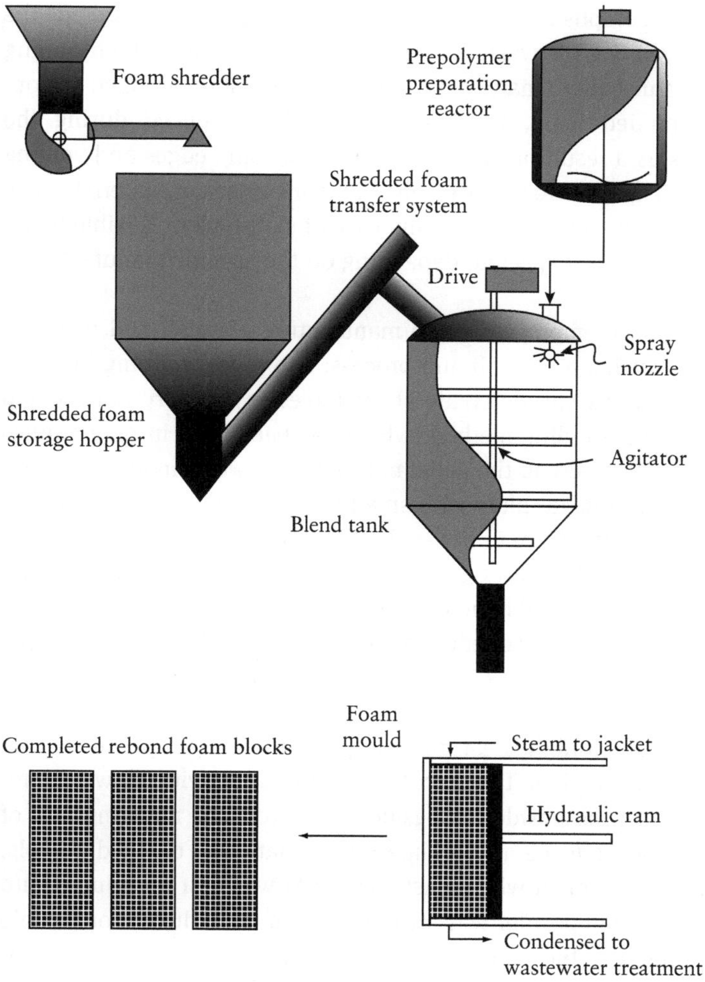

Figure 2.1: Typical rebinding equipment using batch moulding. Reproduced with permission from the European Diisocyanate & Polyol Producers Association, Brussels, Belgium. ©European Diisocyanate & Polyol Producers Association.

a fine powder and (ii) mixing powder with the polyol component to fabricate new PU. The first step necessary for PU recycling involves reduction of the particle size of the original PU product to a level that allows the material to be reprocessed in a secondary production step.

In this recycling method, the optimum particle size is 50–200 µm, depending on the application. Different types of PU waste products, which consist of old recycled parts or production waste (e.g., trimmings, scrap parts), are converted to flake, powder or pellets depending on the PU type that is being recycled. In most cases, this step can be accomplished by shredding or grinding the PU to provide the required media for recycling. This waste is typically a direct result of the trimming of slabstock buns and subsequent manufacturing into the finished product. For operations involving moulded foam, the waste is usually produced during the manufacturing process as a result of the flashing at the closing edges and venting holes of the mould which, in extreme cases, may amount to 6% of material waste depending on the size of the overall part. In the case of slabstock or flexible foam, ≤12% of cutoff waste should be expected, depending on the specific manufacturing process.

Several methods can be employed in the manufacture of small and fine particles. The first option is using a two-roll mill process: two rollers rotating in opposite directions and at various speeds create shear forces in the very narrow gap between them. The second grinding method (which has been used in pilot studies for flexible foam wastes) is based on the pellet mill process: two or more metal rollers press the PUF through a metal plate with small holes or a die. Other technologies (e.g., precision knife cutters) are under evaluation for grinding of flexible PUF. In addition, a process combining cutting and mixing involves a high-shear mixer installed in a polyol tank. The added benefit of this method is prevention of thermal degradation of the powder during size reduction. Glass-filled RIM parts require special grinding processes. The impact disc mill appears to be an appropriate method for pulverisation of such very tough parts [4].

At the pilot-study scale, a high-shear mixer appears to be appropriate for providing the right quality of mixing. The step to operational activity, however, requires suitable metering of the powder alongside the polyol. The metering unit of the entire PU machine needs to be appropriate for the handling of filled polyols. Such technologies were developed when glass-filled RIM was first introduced into the market or when melamine powder was introduced into the industry for flexible PUF. The moisture content of the powder is important, and drying of the powder before mixing may be necessary.

One factor that limits the amount of scrapped PUF that can be recycled by regrinding is the viscosity of the polyol–fine particle mixture that can be handled by the processing equipment. In practice, this can be ≤15 wt% regrind if MDI and a polyol mixture is used, and ≤25% if toluene diisocyanate (TDI) is used as the isocyanate. The main problem of grinding processes for waste processing is that of economy: grinding of scrap PUF to <100–125 μm necessitates more energy consumption. PUF in automotive seating has been recycled using regrind processes. Data obtained from laboratory tests reveal that new moulded foam seats have been fabricated containing 15–20% recycled material and exhibiting excellent processing characteristics. Depending on foam type and filler loading, physical properties may be affected, so adjustment of material specifications may be required (Table 2.1).

The investment cost of first-generation equipment has limited the operational potential of this process to slabstock. Current developments are directed to

Table 2.1: Comparison of the specifications of standard and recycled flexible car seat foams.

Property	Standard	Recycled
Powder content (%)	–	15
Indentation hardness (N)	360	360
Compression set (50%)	5.6	7.0
Tensile strength (kPa)	152	144
Elongation at break	132	103

Reproduced with permission from M.M. Alavi Nikje, A. Bagheri Garmarudi and A.B Idris, *Designed Monomers and Polymers*, 2011, **14**, 5, 395. ©2011, Taylor & Francis [1]

providing economically viable processes to smaller operations. A typical scheme of regrinding technology is shown in Figure 2.2 [1, 4].

Bayer Corporation has invented a process for making energy-absorbent foam with PU wastes. Fillers that produce energy-absorbent foam have properties that are comparable to those prepared from 'virgin' materials.

2.2.3 Adhesive pressing

Adhesive pressing is a rapid and simple recycling method. In this process, the surface of PU particles is coated using an adhesive binder and bonded in a heated press. It has been suggested that adhesive pressing is one of the oldest processes for physical recycling of flexible PUF. During the 1990s, its market potential was ≈10,000–20,000 tonnes per year in Western Europe and 50,000–60,000 tonnes per year in North America. In Europe, RIM particles are being recycled as underfloor heating pipes (which have a very great recycling capacity, especially in Germany) in buildings. A typical method is to mix PUF scrap particles with MDI followed by forming at 100–200 °C under 30–200 bar. This method has enabled useful materials for insulation panels, carpets and furniture to be created [1].

2.2.4 Compression moulding and injection moulding

The compression process involves moulding PU particles at 180 °C and under 350 bar to flow the 'neat' particles together (without any binder) to produce compression moulded parts containing 100% recycled material. Compression moulding is used primarily for RIM PU waste resources, and can be used to manufacture high-performance recycled products. In this procedure, RIM as well as reinforced RIM

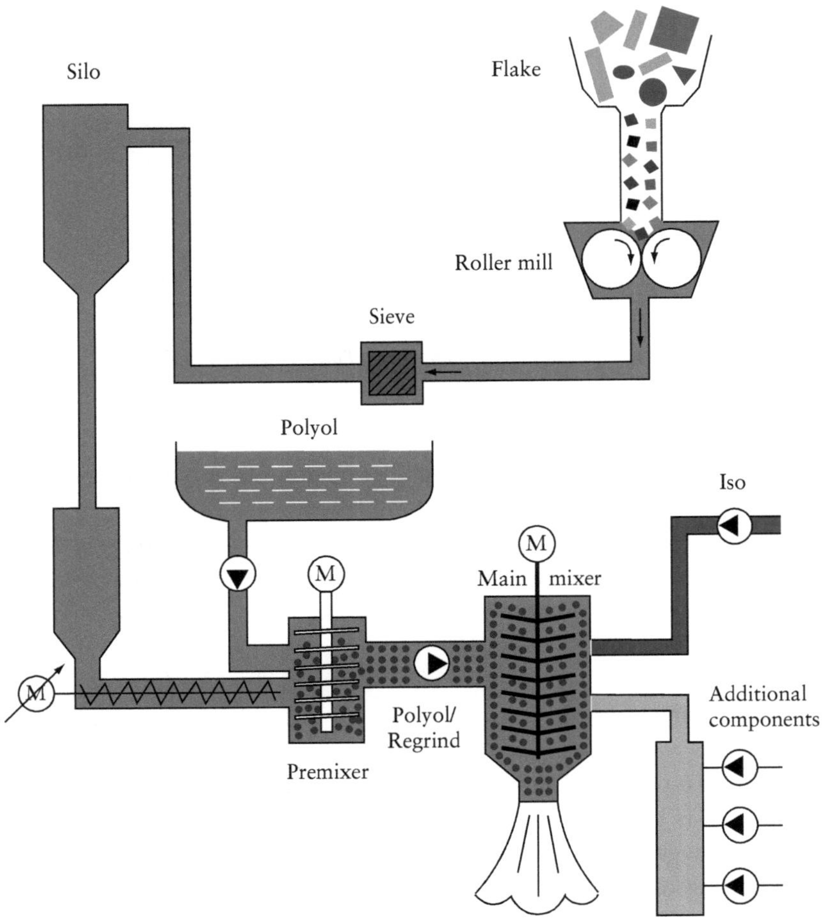

Figure 2.2: Typical regrinding process for PU recycling. Reproduced with permission from the European Diisocyanate & Polyol Producers Association, Brussels, Belgium. ©European Diisocyanate & Polyol Producers Association.

parts are ground into fine particles and subjected to high pressure and heat to produce a solid material that is ideal for many automotive applications.

Similarly, received wastes from crosslinked PU from injection moulded parts can be recycled by this method. This process can also be carried out on a mixture of PU and other plastics because the presence of thermoplastics helps to improve the performance of the final product. A block diagram demonstrating recycling of thermoplastic-containing PU wastes is shown in Figure 2.3.

From a processing viewpoint, compression moulding of material recycling (including a significant portion of PUF) involves conversion of foam into particles sized 0.25–1.0 inch (preferably 0.25–0.50 inch) and transferring these particles to a

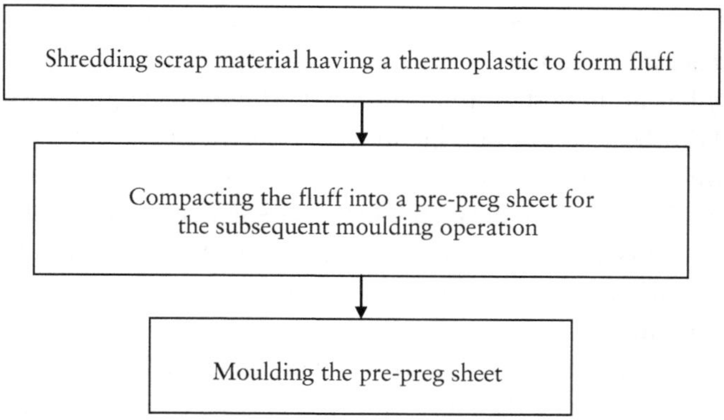

Figure 2.3: Recycling of thermoplastic-containing PU wastes. Reproduced with permission from M.M. Alavi Nikje, A. Bagheri Garmarudi and A.B Idris, *Designed Monomers and Polymers*, 2011, **14**, 5, 395. ©2011, Taylor & Francis [1].

mould and moulding at high pressure and temperature. Preferably, as the glass transition temperature of the PUF is exceeded, other fillers can be mixed with materials that have flexible long chain and easy vibrated moieties. [1]. A block diagram of the compression moulding recycling process is shown in detail in Figure 2.4.

2.2.5 Energy recovery (combustion–incineration)

Combustion (or incineration with the recovery of energy) is an effective way to reduce the volume of organic material that otherwise would have to be sent to a landfill. Combustion is appropriate for all materials for which material recycling is ruled out on ecological or economic grounds or because of logistical difficulties. Rotary kilns, fluidised beds, two-stage incinerators, or mass-burning equipment are appropriate for combustion of plastic scrap.

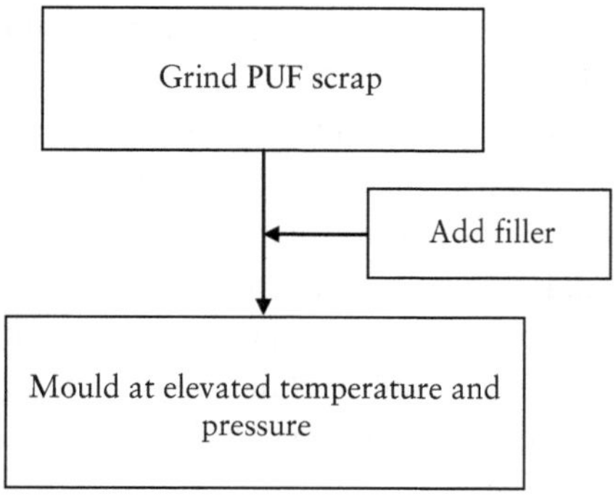

Figure 2.4: Compression moulding for PU recycling. Reproduced with permission from M.M. Alavi Nikje, A. Bagheri Garmarudi and A.B Idris, *Designed Monomers and Polymers*, 2011, **14**, 5, 395. ©2011, Taylor & Francis [1].

For PU recycling *via* this method, energy recovery is the only appropriate disposal process for PU wastes that cannot be marketed or have applications. Incineration with energy recovery has an important role within recycling/reuse options for plastics (including PU). With 86% of the world's oil, coal and gas being burnt for energy, there is a powerful case for partial substitution of energy resources with waste plastics. If it is identified to be the most suitable waste-management option, technology can ensure that emissions from waste-combustion plants are managed safely. This strategy applies strongly to scrap with PU laminates for wood, leather, fabrics, or commingled materials.

Flame retardants within plastic formulations make recovery and recycling very difficult because of their performances upon heating. Recycling offers the most valuable solution if other recovery options do not demonstrate environmental benefits or if they are more expensive. Successful energy recovery schemes depend mostly on the economics of the collection system, with facilities for separation and pretreatment of mixed plastics waste. Based on an efficient logistic framework, energy recovery can become the most cost-effective recovery option. More needs to be done to establish the infrastructure throughout Europe, but incineration with energy recovery is crucial to suitable management of plastic waste (including PU). The relative cost of the different energy-recovery options to society is similar for mono-combustion or for use as co-fuel for cement kilns. However, the cost of collection, sorting, pretreatment and transport might add considerably to the overall cost. Incineration of a PUF results in a volume reduction of ≈99%, which has considerable implications for reduction in landfilling of this material and, simultaneously destroying chlorofluorocarbons and other harmful foam blowing agents. Flame retardants complicate matters, and the effect of these compounds on combustion methods receives much attention.

Due to prevention of depletion of the ozone layer, global warming, greenhouse effects as well as aiming to reduce emissions of carbon monoxide (CO) and nitric oxide (NO), the effect of excess air has been investigated at a combustion temperature of 900–1,000 °C. NO emissions have been found to increase with increasing primary and secondary air flow. Conversion of nitrogen in PU to NO is ≈5.7%, and that of carbon in PU to CO is 0.55%. Further reduction of NO emissions could be obtained with recirculation of flue gases to primary air, which would also complete burnout of traces of hydrocarbons, hydrogen cyanide (HCN) and ammonia (NH_3). In this case, obtained data from two different experiments show that by replacing 50 and 100% of the primary air by the recirculated flue gas led to the reduction of emitted CO and NO gases by 15 and 45%, respectively when compared with data from non-recirculated reaction one.

The behaviour of nitrogen from polymers and plastics in waste-derived fuels during combustion (including PUF) has been studied. It was found that emissions of NO + nitrogen dioxide (NO_2) during combustion in an entrained gas quartz tube reactor (750–950 °C, O_2/N_2 = 7: 93) was dependent upon the amount of char produced from high-nitrogen fuels (PUF, Nylon, urea–formaldehyde glue, sewage sludge) and the nitrogen content of the fuels. At a nitrogen content of 6.6 wt%, <10% of the nitrogen in PUF was emitted as NO + NO_2 [1].

2.3 Thermochemical recycling

In thermochemical recycling of PU wastes, the waste stream is used as a source of energy, fuel or monomers of some value. A schematic flow diagram of all these processes is shown in Figure 2.5 [5].

2.3.1 Pyrolysis

Pyrolysis is a well-known recycling process for PU wastes. Pyrolysis utilises a heated, oxygen-free environment to pyrolyse plastics into gases and a mixture of monomers. In this method, mass loss starts at ≈250 °C and *plateaus* at a mass loss of 80% at 300 °C, whereas the remaining 20% is decomposed until 500 °C. Monitoring of this process is by differential scanning calorimetry, which shows various exothermic peaks during the process, with the most important being at 250–300 °C and others being at the start (340 °C) and end (490 °C) of the second decomposition stages. A yellow smoke and viscous liquid are manufactured in the first stage, followed by slow decomposition of the liquid into gaseous products.

Heating PU samples at 300 K/s under a flow of nitrogen or air (850–1,000 °C) produces considerable amounts of CO, methane (CH_4), HCN, NH_3 and NO in addition to small amounts of ethylene (C_2H_4) and acetylene (C_2H_2). Measured amounts as a fraction of carbon in PUF (for CO and CH_4) and nitrogen (for HCN, NH_3, NO) have been found to be 26, 18, 34, 8 and 21% at 850 °C, and 24, 17, 64, 14 and 9% at 1,000 °C, respectively. Formation of CO and NO is described through intermediates such urethane (HNCOO). Some scholars argue that some nitrogen in PU is released directly as NO, followed by a large release of HCN, which may also be oxidised in the gas phase to NO.

Pyrolysis of a PU adhesive is also a common method. Decomposition of the PU to a mass loss of ≈95% occurs at 230–380 °C. RIM pyrolysis usually gives (at >450 °C) 5–25 wt% char, 10–45 wt% liquids and >40 wt% gases. The product is a red, viscous, single-phase oil with a viscosity that increases with time. To increase the amount, quality and marketability of the liquid, use of activated carbon and PU chars in a secondary pyrolysis reaction step have been tested.

2.3.2 Gasification

Gasification is useful for streams of mixed waste. It consists of an exothermic process that produces heat, ashes and a gas fraction that contains a combustible

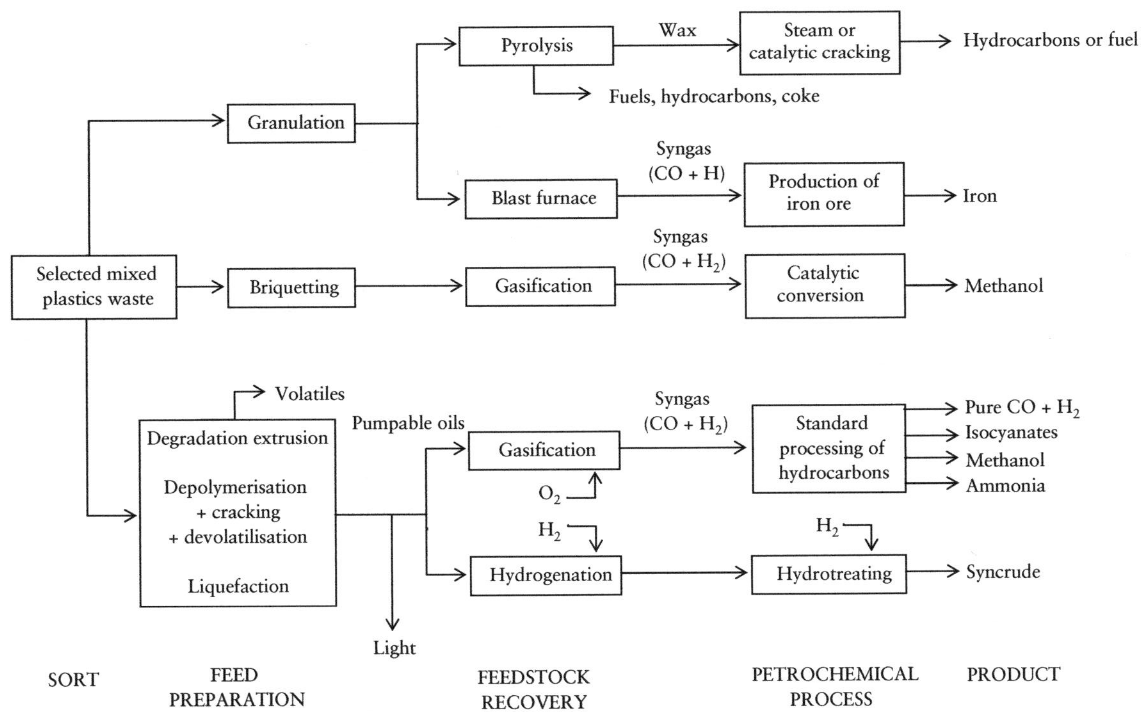

Figure 2.5: Possible chemical products obtained through feedstock thermochemical recycling. Reproduced with permission from the European Diisocyanate & Polyol Producers Association, Brussels, Belgium. ©European Diisocyanate & Polyol Producers Association.

mixture of hydrogen (H_2) and CO ('synthesis gas'). In this method, a waste stream is heated and then combined with air/oxygen (O_2), which forms a CO + H synthesis gas, which can go into refinery processes for production of various chemicals (e.g., NH_3, alcohols). A pump is required for liquid feedstock, which is achieved by liquefaction at a process condition of 20–80-bar of oxygen media and 1,200–1,500 °C. In a few seconds it produces 98–99% conversion into gases plus a slag. The H_2 and CO produced can be also used as raw materials in production of polyether and isocyanates, respectively.

2.3.3 Hydrogenation

Hydrogenation is a 'compromise' between pyrolysis and gasification. The effect of heat and high-pressure H_2 results in gaseous and liquid products. Hydrogenation contains one step more than pyrolysis to enable fabrication of gases and oils of even greater purity through a combination of heat, pressure and hydrogen. The viability of this process requires resolution of two important issues: (i) purity of gases and oils derived from pyrolysis and hydrogenation and (ii) associated costs of fabrication of functional finished products. These products can be used as fuel (partly used as an energy source for the process) and chemical feedstock.

2.4 Chemical recycling

Chemical or feedstock recycling changes the chemistry of the material. The general purpose of chemical recycling of PU is conversion of waste into its original initial raw materials (especially polyols and/or other valued raw materials). That is, the main goal of chemical recycling is recovery of monomers as the main products of the breaking down of urethane bonds under controlled reaction conditions. This goal can be achieved using hydrolysis, aminolysis and glycolysis as well as modified and combined methods (Figure 2.6).

2.4.1 Hydrolysis

Recycling of PU wastes by hydrolysis started during the 1970s in the USA by application of superheated steam at 200 °C. This method converts PUF to a two-phase liquid within ≈15 min in an oxygen-free medium and under pressure.

Hydrolysis of PU waste involves decomposition of the material to its virgin components by treatment with water at elevated temperatures. That is, hydrolysis is a reaction of PU with water that produces polyols and amine-based products. The recycled polyols can be used as monomers in PU formation, and the amine

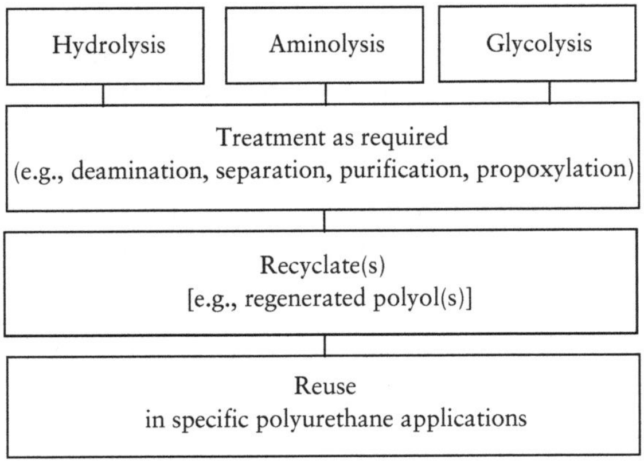

Figure 2.6: Chemical recycling methods for treatment of PU waste. Reproduced with permission from the European Diisocyanate & Polyol Producers Association, Brussels, Belgium. ©European Diisocyanate & Polyol Producers Association.

intermediates can be reused to manufacture other PU components (e.g., isocyanates). The chemistry of a general hydrolysis method is shown in Equation 2.1:

$$\begin{aligned} &R' - NCOO - R'' + H_2O \rightarrow R' - NH_2 + HO - R'' + CO_2 \\ &R' - NH - CO - NH - R'' + H_2O \rightarrow R' - NH_2 + R'' - NH_2 + CO_2 \end{aligned} \tag{2.1}$$

As shown in the reaction formula, monomers are recovered as their previous forms (e.g., polyols) or as functional group interconversion (FGI) forms (e.g., amines as FGI forms of isocyanate functional groups) from breakdown of PU and polyurea bonds, respectively. After recycling, the distilled and purified polyol can be used as a component in new PU formulations. Studies on the recovery and recycling of the polyol demonstrated that the superheated steam temperature should be ≈288 °C, producing a polyol that gave excellent new PUF for seat cushions if mixed at 5% to virgin material. Too high a temperature gives a decrease in recovery of useful polyols, whereas polyols recovered at too low a steam temperature produces an unstable new PUF due to unsuccessful recovery of the polyol. PU hydrolysis suffers from unfavourable economics and lack of markets for recycled products.

PU wastes may be converted, in a convenient and economical way, to useful active hydrogen-containing polyethers and polyamines by a reaction of the PU with water, strong base, and an activating agent such as a quaternary ammonium compound containing ≥15 carbon atoms or an organic sulfonate containing ≥7 carbon atoms. The activating agent acts as phase-transfer catalyst and helps to accelerate the rate of hydrolysis and monomer recovery at high yields and purities.

Recent studies have shown recovery of other products besides polyols, and that alkaline compounds may accelerate the hydrolysis. Kabushiki Kaisha and Takeda Chemical Industries in Japan have invented an apparatus for hydrolysis of polyisocyanate derivatives having at least one isocyanate group that can be hydrolysed into raw materials or derivatives thereof for target compounds. This apparatus comprises a hydrolysis section for bringing only pressurised water in a liquid state at 190–370 °C into contact with the target compounds in the reactor and a post-processor for conducting dewatering, addition, distillation, separation, and liquid separation for the hydrolysis product discharged from the reactor [6].

The reaction of polyether-based PUF with dry atmospheric pressure steam can also be used to recover waste PUF. Completion of this reaction at 190–230 °C yields high-quality polyols, isomeric toluene diamines (TDA), and CO_2 as primary products. Kinetic analyses of TDA yield *versus* time suggest a parallel first-order reaction scheme wherein urethane linkages react ≤50× faster than urea linkages. Results derived from kinetic investigations have demonstrated that urethane linkages are broken by direct hydrolysis, whereas urea linkages undergo thermal dissociation to the parent isocyanate and amine.

Hydrolysis of PUF and recovery of diamines and polyethers (or polyesters) from the hydrolysate would be carried out near atmospheric pressure and temperatures >185 °C. The reaction zone would contain water vapour, which involves increasing the rate of diamine production by inclusion of a minor (but effective amount) of NH_3. Volatilised diamines are separated from the foam and removed from the reaction zone (e.g., in overhead stream) whereas polyethers (or polyesters) are recovered [7].

2.4.2 Aminolysis

Aminolysis is the catalysed reaction of PU waste with amines such as dibutylamine, ethanolamine (EA), lactam or lactam adducts under pressure and at high temperature. The well-known metal hydroxides potassium hydroxide (KOH) and sodium hydroxide (NaOH) are the main catalysts, and polyols, carbamates and aromatic amines are the main products. To improve performances of the products, ethylene or propylene oxide (PO) is reacted with the amines, which leads to the formation of amine-based polyols. Finally, the separated polyols can be used in the new PUF formulation alone or in combination with a virgin polyol. Considerable changes are observed in the properties of the new foam if the amine contaminants do not separate from the recycled polyol.

For example, MDI-formulated flexible PUF are recycled by alkanolamines without a catalyst at 150 °C to recover polyol monomers and amine-based compounds. The decomposition product is completely separated into two phases. The upper liquid-phase is amine-contaminated polyether polyols, and the lower liquid-phase is 4,4′-methylenedianiline (MDA) and alkanolamines derivatives. Chuayjuljit and coworkers [8] reported the possibility of carrying out aminolysis reactions of rigid PUF scraps using diethylenetriamine (DETA) as a degradation reagent and NaOH as the catalyst.

In general, this reaction was done at 180 °C with a rigid PUF:DETA ratio of 2:1 and yielded a polyol and MDA as the reaction products. Chuayjuljit and colleagues found that a higher degree of depolymerisation increased with an increasing amount of NaOH. In contrast, the amount of MDA decreased due to the aminolysis of PUF by MDA catalysed with NaOH.

In another work, flexible PUF decomposed into a polyol mixture using an extruder with diethanolamine (DEA) as a decomposing agent. The resulting decomposed product could be employed as an alternative virgin polyol for reclaiming PU. In the case of reclaiming an elastomer using the decomposed product without purification, virgin polyol could be alternated with the decomposed product at ≤80%. It is a great improvement compared with reclamation using foam, the percentage of which was a maximum of 5%. Furthermore, the percentage could be improved to ≤100% by purification of the decomposed product [9].

2.4.3 Glycolysis

Glycolysis is the most widely used chemical recycling process for PU wastes. This process was developed during the 1980s. Figures 2.7 and 2.8 show the schematic mechanism of the glycolysis process. The main aim of this method is recovery of the valued monomers (polyols from waste PU) for the production of new material [1].

The glycolysis process implies heating ground PU scraps at ≤180–220 °C in high-boiling point glycols in the presence of a catalyst. Glycols act as destroying agents by attaching to urethane functional groups, followed by breaking of bonds and liberation of amines and polyols.

Different methods have been reported in chemical recycling, all of which have merits and drawbacks. The destroying agent typically selected is DEG and the co-reagent is an amine, DEA. Carrying out the reaction at >220 °C gives undesired side-reactions towards amines and, at <180 °C, gives too-low catalyst activity. One of the most important drawbacks in using chemical reagents is applying amine as a co-reactant, which causes problems in using end-processed and recovered materials. Amines act as catalysts and enhance the reaction rate between polyol and polyisocyanate by increasing the electrophilicity of the isocyanate functional group and facilitate attachment of polyols and fabrication of PU in too-short and uncontrollable reaction conditions. Hence, after treatment of PU wastes by destroying agents containing amine, amine content in the polyol portion must be lowered to prevent undesired reactions and to find controllable models of reaction kinetics.

Two methods can be employed wherein a single polyol is recovered or flexible and rigid polyol components are recovered [10–12]. A method for preparing a recyclate

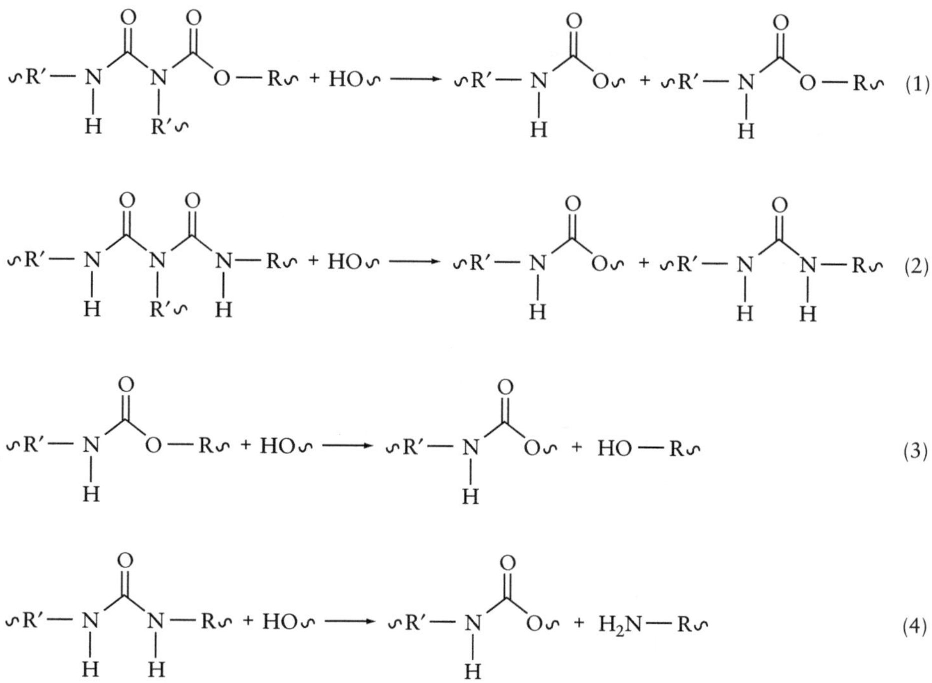

Figure 2.7: Glycolysis of urethane functional groups and their co-existing functional groups such as allophanate, biuret and urea. Reproduced with permission from M.M. Alavi Nikje, A. Bagheri Garmarudi and A.B Idris, *Designed Monomers and Polymers*, 2011, **14**, 5, 395. ©2011, Taylor & Francis [1].

polyol obtained from glycolysis of PU comprises reacting scrap PU with a short-chain compound containing ≥2 hydroxyl groups in the presence of a catalyst and by adding a cyclic carbonate. The method results in a recyclate polyol having a low content of primary aromatic amine, which may be used to manufacture cellular or non-cellular PUF, polyurea-PU, or polyisocyanurates [13].

As mentioned above, the main goal in monomer recovery is reusing (alone or in blend) by virgin polyol in the new PUF formulation. Recycling and recovery of polyols using a 'foam to foam' process omits waste collections and solves transportation and logistics problems. In the chemicals market, various companies exhibit their recovered polyols from wastes alone and/or in blend with a virgin polyol.

An example of this method is the glycolysis process developed by Getzner Werkstoffe in Austria (Figure 2.9).

A method for double-phase recovery of polyols was developed by ICI Polyurethanes and Du Vergier in Hoddesdon, Hertfordshire, UK, and is referred to as the 'split-phase glycolysis method'. In this method, scrap PUF (preferably based on MDI) is reacted with DEG, produces two product phases in the

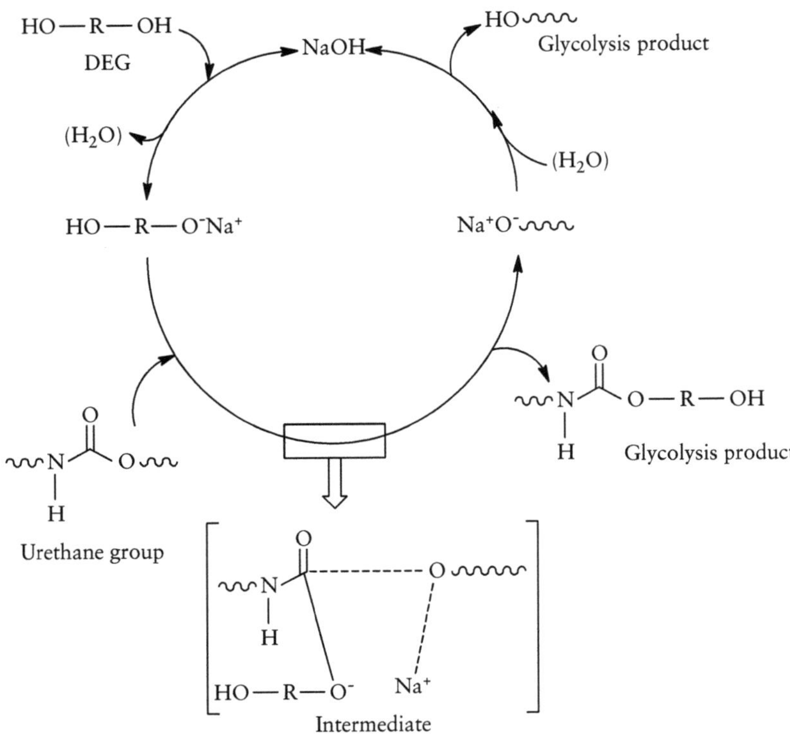

Figure 2.8: PUF glycolysis in the presence of a glycol as a destroying agent and NaOH as the catalyst (schematic). DEG: diethylene glycol. Reproduced with permission from M.M. Alavi Nikje, A. Bagheri Garmarudi and A.B Idris, *Designed Monomers and Polymers*, 2011, **14**, 5, 395. ©2011, Taylor & Francis [1].

reactor or reaction vessel. The low-density layer contains the flexible polyol, and the heavier layer contains MDI-derived compounds, which are converted into a rigid polyol using PO.

Reacting the glycolysis polyol product with an alkylene oxide in a ratio of <1 mol of alkylene oxide per 1 mol of active amine hydrogen atom in the glycolysis polyol product yields a beneficial polyol. It has been shown that in the presence of MDI, both of the split-phases of the glycolysis mixture can be used as components of urethane adhesive mixtures for bonding wood, metals, plastics, glass, ceramics, paper and leather. To recycle the mixture of PU and other plastics, the plastic waste is reacted with an alcohol mixture comprising at least one low-molecular weight (Mw) bifunctional alcohol and at least one polyether polyol. Then, the resultant glycolysis polyol is separated from the other plastic waste. Recently, several research teams have investigated novel glycols, new methodologies for chemical treatment of PU waste, and application of the recycled polyols in new PUF formulations.

BASF has invested resources and technology to develop what are believed to be practical methods for the chemical recycling of PU-process waste and end-of-life

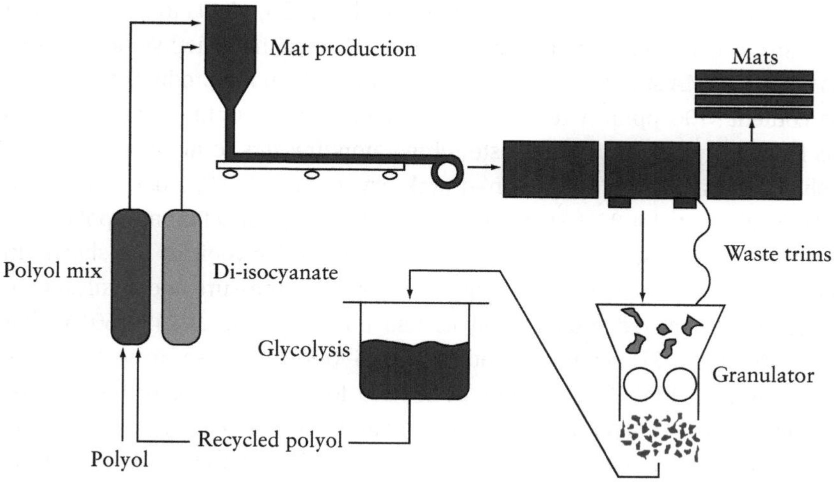

Figure 2.9: The glycolysis concept as developed by Getzner Werkstoffe in Austria. Reproduced with permission from the European Diisocyanate & Polyol Producers Association, Brussels, Belgium. ©European Diisocyanate & Polyol Producers Association.

waste. Initially, the well-established glycolysis process was developed at BASF (Schwarzheide, Germany). BASF's Laboratory in Wyandotte (MI, USA) has continued research and development in this area, concentrating on modifying the chemistry and process to suit their PU customers in North America. BASF's second-generation chemistry has resulted in homogeneous recycled polyols with residual amine content <0.1% in a one-pot method operating at moderate temperatures and ambient pressure.

Another method for the production of recyclate polyols invented by BASF involves the reaction of rigid PUF with short-chain hydroxyl-containing compounds. This reaction comprises addition of epoxidised native fatty oils to the reaction mixture after glycolysis at 10–80 °C below the glycolysis temperature.

Properties of the recyclate are usually dependent on the properties of the stream of PU waste. A glycolysate from high-density rigid non- cellular PU can be made using second-generation technology. It can be reformulated into various rigid and semi-rigid foams at ≤100% polyol loading while providing supreme mechanical and physical properties in the finished product. Examples include polyisocyanurate foam, structural foam, insulation non-critical foam, energy-absorbing foam, low-density void-filling foam, elastomeric coatings, non-cellular rigid PU, and moulded semi-flexible foam. Members of the European Diisocyanate & Polyol Producers Association and other research teams have optimised single-phase glycolysis. The split-phase glycolysis process, in which the product separates in two phases, has been developed up to pilot scale for MDI flexible foams. The viability of the glycolysis process appears to be in recycling of production waste as opposed to post-consumer waste.

The glycolysis of water-blown PUF by ethylene glycol (EG) yields the polyol and a solution of ureas, carbamates and amines in the EG, then hexamethylenetetramine (HMTA) is added. HTMA suppresses creation of solid-phases in the products at a diaminotoluene content <100 ppm. A research team in Taiwan has optimised the process conditions for glycolysis of PU from waste refrigerators/freezers to manufacture high-quality polyol recyclate [14–17]. In a stirred tank reactor at ≈220 °C, 1 bar, and a residence of 2 h is adequate for 87–95% recovery with DEG as the reactant and potassium acetate as the catalyst. A second study on PUF from end-of-life vehicles has shown an optimal residence time of 90 min at the same temperature, pressure and catalyst conditions. Then, the recovered product polyols had boiling points of 245–260 °C. For both studies, the initial PU scrap size had been 0.15–0.85 mm. Also, recycled polyol from the glycolysis of RIM PU scrap has been used to replace ≤60% of the original polyol for new RIM PU final products in Germany. Two-step chemical recycling methods are more convenient than that in the first step of the method. PU-containing scrap is subjected to a glycolysis reaction to produce polyol products. In the second step, the recycled polyols are used as initiators in a reaction with alkylene oxide to manufacture oxyalkylated polyols for preparation of PU [18].

Simon and co-workers found that the stannous octoate is a suitable catalyst in the glycolysis of PU wastes, yielding a great-quality recovered polyol in a short reaction time [19].

Borda and co-workers [20] investigated the glycolysis of elastomers based on polyether polyol and MDI. Borda and colleagues employed EG, 1,2-propylene glycol, triethylene glycol and polyethylene glycol (PEG) as the glycolysis reagents and DEA as the catalyst. The reaction temperature described was 170–180 °C, with a ratio of glycolysis reagent:PU waste of 2:1 and ratio of DEA:EG between 1:1 and 1:9. The result was split-phase mixture in which the upper liquid-phase was characterised as the starting polyol.

Wang and co-workers [21] reported the glycolysis of thermoplastic polyurethane (TPU) elastomers based on MDI and a polyether polyol (PEG). Obtained data revealed that a ternary mixture containing DEG, EG and EA could be used as degradation reagent for TPU glycolysis. In that study, when DEG and EG were employed as low-Mw glycols, whereas EA and lithium acetate were employed as catalysts, TPU was converted into liquid after treatment with this degradation reagent at 160–190 °C. Wang and colleagues stated that, by using a 9:9:2 ratio of DEG:EG:EA, decomposition products were separated completely two layers, whereby the upper liquid layer was the polyether polyol.

Morooka and co-workers [22] described the glycolysis of received rigid PUF wastes from refrigerators using DEG as the solvent and barium oxide (BaO) or DEA as catalysts. They found that ≤10% wt of the glycolysate obtained could be added to raw virgin polyol to prepare a new rigid PUF with similar thermal conductivities and compressive strengths when compared with the properties of 'pristine' foam.

To investigate the glycolysis potential of PUF, Murai and co-workers [23] studied the dissolution time of rigid PUF under various glycolysis conditions with a simple testing method. They employed dipropylene glycol and tetraethylene glycol as glycols, and KOH and dibutyltindilaurate as catalysts at a reaction temperature of 170–200 °C. This research team found that the reaction temperature, catalyst concentration, and particle size of PUF had a significant effect upon the dissolution time of PUF. Moreover, type of glycol (polyol), its Mw, and type of catalyst affected the dissolution process. They also investigated the influence of the ratio of glycol:PUF weight on the extent of glycolysis of rigid PUF. An increase in the ratio of glycol:PUF weight from 200:150 to 200:100 enhanced the extent of the glycolysis reaction.

Wu and co-workers [24] studied the glycolysis reactions of flexible PUF under atmospheric pressure and isothermal condition at 220 °C using DEG and KAc as the solvent and catalyst, respectively. In that work, the effects of reaction times as well as concentrations of DEG and KAc in the mixture on the properties of glycolysis products were investigated and the optimum reaction condition obtained. They found the optimal concentrations of DEG and KAc to be 150 and 1 wt% of the flexible PUF, and that the glycolysis reaction was concluded when the reaction time was 90 min.

Datta and co-workers [25] investigated the glycolysis reaction of elastic PUF using 1,6-hexanediol (1,6-HD) as a low-Mw glycol and KAc as the catalyst. They carried out glycolysis reactions at 230–245 °C at a ratio of PUF:1,6-HD mass of 1:1, 2:1, 4:1, 6:1, 8:1 and 10:1, respectively, by applying the catalyst as 0.5% by weight with respect to the PUF mass. They observed that at PUF:1,6-HD ratios of 1:1, 2:1 and 4:1, a single-phase liquid product was obtained. On the other hand, the reaction was run as a 'split-phase' glycolysis when ratios of PUF:1,6-HD of 6:1, 8:1 and 10:1 were used. They also observed that the reaction time was a function of the ratio of PUF foam:1,6-HD mass. Furthermore, the obtained products with the highest ratios of PUF:1,6-HD mass were used as the only polyol components in the new formulations of PU elastomers. Datta [26] investigated the influence of different low-Mw glycols on chemical structure and thermal stability of glycolysates as PU intermediates. In that work, the intermediates were obtained by the glycolysis process of waste PUF in the reaction with different glycols such as EG, 1,3-propanediol, 1,4-butanediol, 1,5-pentanediol and 1,6-HD. Fourier-Transform infrared analyses showed the presence of similar chemical substances in glycolysates irrespective of the glycolysis agent used. Results of thermogravimetry analyses demonstrated that the thermal resistance of glycolysates increases with the increasing Mw of glycol used in glycolysis.

Molero and co-workers reported the glycolysis of flexible PUF in the split-phase with different low-Mw glycols to study their activity and select a system to obtain the highest-quality recovered polyol.

They found that the time required to reach complete conversion, the chemical properties of the polyol phase, and its purity was dependent upon the glycol employed. DEG proved to be the most appropriate glycol to obtain high purity in the polyol phase [27].

Molero and co-workers [28] described a purification process of recovered polyols. It consisted of a liquid extraction with a slightly acid pH solvent. They found that ≥60 °C and a minimum solvent/upper-phase ratio of 1:1 provided a purified polyol that could undergo foaming [29].

Simón and co-workers [30] developed a process for the 'valorisation' of the glycolysis bottom-phase. It consisted of recycling the excess of glycol employed in the split-phase glycolysis of PU wastes by distillation under reduced pressure (50 mbar). On the other hand, the residue of the distillation column (which contained the isocyanate part of the byproducts of glycolysis) was assayed as an initiator in the synthesis of new polyols. Thus, this residue of the distillation column was employed as an initiator of a rigid polyol, and substituted for the commonly used commercial one based on TDA.

Zhu and co-workers [31] reported the glycolysis of the waste rigid PUF from waste refrigerators. In that study, the roles of glycolysis reagents (EG and DEG) and catalysts [NaOH, sodium acetate (NaAc) and triethanolamine] were investigated according to the mechanism of glycolysis. Results showed that the glycolysis efficiency of EG was higher than that of DEG, and that the catalytic efficiency of alkali metal salts (NaOH) was better than that of triethanolamine and organic salts of an alkali metal (NaAc).

Santos and co-workers reported the recovery of the polyol from flexible PUF and use of different catalysts in the glycolysis process or depolymerisation reaction. Results showed that metallic catalysts showed higher efficiency than conventional amine catalysts [32].

Molero and co-workers reported the glycolysis of flexible PUF in the presence of various catalysts to obtain high-quality recovered polyols *via* split-phase in a short reaction time. DEA, titanium n-butoxide as well as octoate salts, which are novel compounds for this application, showed appropriate catalytic activity [33].

In another report, Molero and co-workers [34] reported the two-phase glycolysis reactions of flexible PUF with DEG in the presence of stannous octoate as a novel catalyst. Results showed that stannous octoate yielded the highest quality of recovered polyol in a short reaction time.

2.5 Other methods of chemical recycling

Hydroglycolysis is a complex process undertaken by adding water to a glycolysis process in which lithium hydroxide (LiOH) is used as the catalyst. Ford Motor Company has developed this method and claims that it produces a simpler and more valuable product mixture. It is more expensive than conventional glycolysis because the purification of the polyol-rich product is more complicated. However, it has the advantage that complex mixtures of 'dirty' and contaminated PU

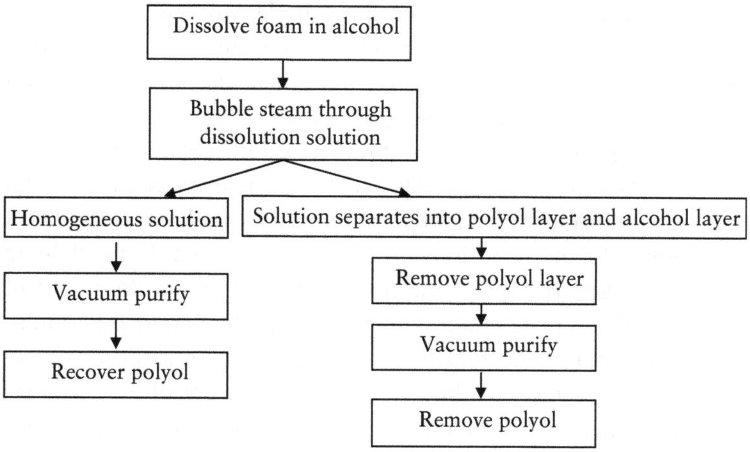

Figure 2.10: Block diagram of a hydroglycolysis recycling method. Reproduced with permission from M.M. Alavi Nikje, A. Bagheri Garmarudi and A.B Idris, *Designed Monomers and Polymers*, 2011, **14**, 5, 395. ©2011, Taylor & Francis [1].

wastes can also be processed, which otherwise would have to be landfilled. The recovered polyol may replace ≤50% of virgin polyol material for PUF. Extraction of pure polyether triols from the products is accomplished using hexadecanes or hydrocarbons with a similar boiling point. A schematic version of the hydroglycolysis method is shown in Figure 2.10 [1].

Another method of chemical recycling is the reaction of waste PUF with alcohol under pressure and at an elevated temperature. This process produces the original polyols together with urethane products. Then, the different components are separated to allow their reprocessing and reuse. This process uses a heated, oxygen-free environment to breakdown PU and plastics into gases, oils and solids, but this method seems difficult to scale up. A new recycling process has been introduced based on a polar and aprotic solvent (e.g., dimethylsulfoxide, *N*-methyl pyrrolidone, dimethylformamide, acetonitrile, tetrahydrofuran, hexamethylphosphoric acid triamide) and mixtures for PU dissolution. In this method, PU waste is added to the polar aprotic solvent, and then a non-solvent is added to the PU solution to prepare a suspension of PU in the mixture. In the next step, the solvent is removed to prepare a suspension of PU in a non-solvent and the filtrate collected for further treatment.

2.6 Current and future developments

Increasing waste-to-energy and other thermal processing activities involving gasification, pyrolysis and two-stage combustion (removing problematic components in the first stage) also allows for the disposal of considerable amounts of scrap PU without difficulties. This method may be slowed down by lack of support for technology development: the current focus is very much on renewable energy (e.g., biomass), and more efficient fossil fuel-derived power and heat instead. Recently, an elastic permeable pavement was fabricated that uses waste PU chips obtained from shoe soles, toys, vehicles, refrigerators and decrepit recycled PU-resilient pavement. This method solves the odour problem of waste tyres and prevents environmental pollution, and provides comfort and shock absorption during walking. Addition of 'micronised' PU powders obtained from flexible PU wastes to urea–formaldehyde resins and to phenol–formaldehyde resins improves markedly the performance of the panels produced with these resins (i.e., plywood and particleboard). Formulations of adhesives and film materials from PU waste suggest a versatile resource-saving and environmentally friendly method for reuse of polyglycol adipate urethanes with microcellular and dense structures. This concept allows transformation of a polymer waste into a reclaim with an elevated level of service parameters. Various methods for material and chemical recycling of PU materials have been developed, by far the most important being glycolysis and regrinding. New trends in chemical recycling of PU waste have focused on two goals: new recycling agents and novel sources for process activation. Nowadays, recycling is attempted using PU waste without regard to the separation or selection of waste. Laboratory-based glycolysis processes have been carried out using two types of PUF recovered from shredder residue:

- Best-case scenario: handpicked PUF from shredder residue with >80% conversion into a liquid initiator.
- Worst-case scenario: mixed PU materials separated by an automated separation method from shredder residue with 50% conversion into a liquid initiator.

TDI- and polydiphenylmethane diisocyanate-based flexible foams are produced from the new recycled polyols prepared by propoxylation of the glycolysed products (initiators) obtained from mixed PU materials. Novel chemical agents are being researched for PU recycling. Microporous PU elastomers have been recycled by *tris* (1-methyl-2-chloroethyl)phosphate-induced degradation. The phosphorus-containing degradation products are transformed into reactive intermediates by reaction with PO and are used for preparation of rigid PUF. Split-phase glycolysis of isocyanate-derived cold-cure PUF is done with glycerin as a new glycolysing agent. Among material recycling methods, this one has been reported to be simple and economically attractive. Glycerin has been used as a 'green' destroying solvent and NaOH as the catalyst in the chemical recycling of PUF wastes [35]. Results have shown that this process can be used to recover monomers in high yield and purity. Spectroscopy methods have been used to elucidate chemical structures and data compared with authentic sample data. Chemical structures were fully adaptable with the structures of commercial polyols, and the foams produced had the same properties as foams obtained from the virgin polyol. The merit of this process was replacement of DEG with glycerin as a green and safe destroying agent. Microwave irradiation has been introduced as a new energy source for the glycolysis method [36–38]. Results have shown that this process is successful for polyol recovery and for use of recycled polyol in the new PUF formulation. Data were comparable with foams produced by virgin polyols. The energy source would reduce the time required for a chemical-recycling method to be undertaken, yielding high-quality products. The number of studies on PU recycling is increasing because of the lack of economic efforts for recyclate marketing. Until reasonable-sized markets for PU waste materials can be created, the final destination for most PU wastes will be landfill until this process is halted by legislation. Due to increasing prices for petrol-based chemicals, using new technologies and methodologies in chemical recovery of the monomers from polymeric wastes could aid protection of the environment and enable development of environmentally friendly methods.

References

1. M.M. Alavi Nikje, A. Bagheri Garmarudi and A.B Idris, *Designed Monomers and Polymers*, 2011, **14**, 5, 395.
2. R. Zevenhoven in *Treatment and Disposal of Polyurethane Wastes: Options for Recovery and Recycling*, Energy Engineering and Environmental Protection Publications, Helsinki University of Technology Department of Mechanical Engineering, Espoo, Finland, 2004.
3. *Recycling and Recovering Polyurethanes: Rebonded Flexible Foam*, Fact Sheet, European Diisocyanate & Polyol Producers Association (ISOPA), Brussels, Belgium. http://www.isopa.org/media/1104/isopa-factsheet_recycling_and_recoverypolyurethanes_v02.pdf.
4. *Recycling and Recovering Polyurethanes: Regrind/ Powdering*, Fact Sheet, European Diisocyanate & Polyol Producers Association (ISOPA), Brussels, Belgium. http://www.polyurethanes.org/uploads/documents/regrind.pdf.
5. *Recycling and Recovering Polyurethanes: Feedstock Recycling*, Fact Sheet, European Diisocyanate & Polyol Producers Association (ISOPA), Brussels, Belgium. http://www.polyurethanes.org/uploads/documents/feedstock.pdf.
6. J.S. Robert and C.D. Clifford, inventors; General Motors Corporation, assignee; US4328368, 1982.
7. S. Mishra, V.S. Zope and R.D. Kulkarni, *Polymer–Plastics Technology and Engineering*, 2004, **43**, 4, 1001.
8. S. Chuayjuljit, C. Norakankorn and V. Pimpan, *Journal of Metals, Materials and Minerals*, 2002, **12**, 1, 19.
9. T. Fukaya, H. Watando, S. Fujieda, S. Saya, C.M. Thai and M. Yamamoto, *Polymer Degradation and Stability*, 2006, **91**, 11, 2549.
10. M.M. Alavi Nikje, M. Haghshenas and A. Bagheri Garmarudi, *Polymer–Plastics Technology and Engineering*, 2006, **45**, 4, 569.
11. C. Molero, A. de Lucas, F. Romero and J.F. Rodríguez, *Journal of Applied Polymer Science*, 2008, **109**, 1, 617.
12. M.M. Alavi Nikje and F.A. Mohammadi, *Polimery Journal*, 2009, **54**, 7–8, 541.
13. C.H. Wu, C.Y. Chang, C.M. Cheng and H.C. Huang, *Polymer Degradation and Stability*, 2003, **80**, 1, 103.

14. M.M. Alavi Nikje, M. Haghshenas, A. Bagheri Garmarudi, N. Barzandeh and K. Taheri, *International Journal of Pure and Applied Chemistry*, 2006, **1**, 3, 411.
15. J. Datta, *Journal of Elastomers and Plastics*, 2010, **42**, 2, 117.
16. C.H. Wu, C.Y. Chang and J.K. Li, *Polymer Degradation and Stability*, 2002, **75**, 3, 413.
17. H. Benes, J. Rosner, P. Holler, H. Synkova, J. Kotek and Z. Horak, *Polymers for Advanced Technologies*, 2007, **18**, 2, 149.
18. C. Molero, A. de Lucas, F. Romero and J.F. Rodríguez, *Journal of Material Cycles and Waste Management*, 2009, **11**, 2, 130.
19. D. Simón, M.T. García, A. de Lucas, A.M. Borreguero and J.F. Rodríguez, *Polymer Degradation and Stability*, 2013, **98**, 1, 144.
20. J. Borda, G. Pässztor and M. Zsuga, *Polymer Degradation and Stability*, 2000, **68**, 3, 419.
21. X. Wang, H. Chen, C. Chen and H. Li, *Fibers and Polymers*, 2011, **12**, 7, 857.
22. H. Morooka, T. Nakakawaji, S. Okamoto, K. Araki and E. Yamada, *Polymer Preprints*, 2005, **54**, 1, 1951.
23. M. Murai, M. Sanou, T. Fujimoto and F. Baba, *Journal of Cellular Plastics*, 2003, **39**, 1, 15.
24. J. Datta and M. Rohn, *Journal of Thermal Analysis and Calorimetry*, 2007, **88**, 2, 437.
25. J. Datta, *Journal of Thermal Analysis and Calorimetry*, 2012, **109**, 2, 517.
26. C. Molero, A. de Lucas and J.F. Rodríguez, *Polymer Degradation and Stability*, 2006, **91**, 2, 221.
27. C. Molero, A. de Lucas and J.F. Rodríguez, *Solvent Extraction and Ion Exchange*, 2006, **24**, 5, 719.
28. D. Simón, A.M. Borreguero, A. de Lucas, C. Molero and J.F. Rodríguez, *Journal of Material Cycles and Waste Management*, 2014, **16**, 3, 525.
29. P. Zhu, Z.B. Cao, Y. Chen, X.J. Zhang, G.R. Qian, Y.L. Chu and M. Zhou, *Environmental Technology*, 2014, **35**, 21, 2676.
30. L.M. dos Santos, C.L.P Carone, J. Dullius, R. Ligabue and S. Einloft, *Polímeros*, 2013, **23**, 5, 608.
31. C. Molero, A. de Lucas and J.F. Rodríguez, *Polymer Degradation and Stability*, 2006, **91**, 4, 894.
32. M.M. Alavi Nikje and M. Nikrah, *Polymer Bulletin*, 2007, **58**, 2, 411.
33. M.M. Alavi Nikje and M. Nikrah, *Polymer–Plastics Technology and Engineering*, 2007, **46**, 4, 409.
34. M.M. Alavi Nikje and M. Nikrah, *Journal of Macromolecular Science, Part A: Pure and Applied Chemistry*, 2007, **44**, 6, 613.
35. M.M. Alavi Nikje, M. Nikrah and M. Haghshenas, *Polymer Bulletin*, 2007, **59**, 1, 91.
36. M.M. Alavi Nikje, M. Nikrah and F.H. Agha Mohammadi, *Journal of Cellular Plastics*, 2008, **44**, 5, 367.
37. M.M. Alavi Nikje and F.H. Agha Mohammadi, *Polymer–Plastics Technology and Engineering*, 2010, **49**, 8, 818.
38. M.M. Alavi Nikje in *Proceedings of the 2nd International IUPAC Conference on Green Chemistry*, 14–19th September, St. Petersburg, Russia, 2008.

3 Chemical recycling of flexible and semi-flexible polyurethane foams

3.1 Glycolysis of integral skin foam wastes received from a production line of steering wheels and arm rests

3.1.1 Introduction and background

Because of industrial use of polyurethane(s) (PU) integral skin foams, the first focus is on chemical recycling on foam wastes received from production lines of steering wheels and arm rests. New styling trends, surface design, safety functions, and comfort are made possible by application of integral skin polyurethane foams (PUF). The interior of cars and public transportation vehicles (e.g., bus, rail, airplanes) must satisfy functional and passive safety requirements (e.g., crash energy should be absorbed reversibly by the protection padding). Steering wheels produced with integral skin foam have found wide application in the automotive industry. A large part of the raw materal becomes waste during production (e.g., cutoffs, mould vents). Hence, the cost of taking this material and purchasing base material will be reduced while the same manufacturing output is maintained. There is increasing public awareness that the world has limited natural resources and a limited capacity to manage the volume of waste that human activities generate.

'Competitive' materials are recycled, so alternative uses of trim foam are required to satisfy the industry's requirements for maximum environmental protection and performance. The main purpose of this work is finding a simple and convenient method for recovering used polyols in the foam formulation *via* chemical recycling. The best process may vary from case to case and depends on PU properties, the intended application, the related capacity of the market for the recycled material and, above all, on logistical, economic, and environmental factors.

The glycolysis of PUF wastes into raw materials is one of the principal processes for its recycling. Several studies have indicated that appropriate glycolysis can resolve the disposal problems of waste PU and help to obtain high-quality polyols.

The aim of this study focuses on application of optimum glycolysis conditions for treatment of PU integral skin foams from steering wheels to define an economical industrial process that would be applicable in every situation and would not require special pre- or post-curing of the product [1, 2].

3.1.2 General procedure of recycling

As a general procedure, we carried out a recycling process on common products of integral skin foam formulations in Iran. In this study, the polyether polyol was

https://doi.org/10.1515/9783110641592-003

Daltorim® EK 20,350, the 4,4′-diphenylmethane diisocyanate (MDI) was SUPRASEC® 2082, and additives were used for PU integral skin foam formulations. Technical data for virgin polyol (Daltorim® EK 20,350) and MDI (SUPRASEC® 2082) are listed in Tables 3.1 and 3.2, respectively. The prefix 'virgin' is used to differentiate the polyol (Daltorim® EK 20,350) from recycled polyol. Diethanolamine (DEA), diethylene glycol (DEG) and all catalysts were of industrial grade and purchased from the Iranian Petrochemical Company, and were used as received without further purification. Spectroscopy-grade tetrahydrofuran (THF) and polystyrene standards for gel permeation chromatography (GPC) were purchased from Merck.

Table 3.1: Specification of virgin polyol (Daltorim® EK 20,350).

Appearance	**White viscous liquid**
Viscosity at 25 °C	1,740 mPa.s
Specific gravity at 25 °C	1.025 g/cm^3
Flash point (°C)	265
Fire point (°C)	290

Reproduced with permission from M.M. Alavi Nikje, M. Haghshenas and A. Bagheri Garmarudi, *Polymer-Plastics Technology and Engineering*, 2006, **45**, 4, 569. ©2006, Taylor & Francis [1]

Table 3.2: Specification of MDI (SUPRASEC® 2082).

Appearance	**Brown liquid**
Viscosity at 25 °C	340 mPa.s
Specific gravity at 25 °C	1.23 g/cm^3
Print point (°C)	204
Burning point (°C)	220

Reproduced with permission from M.M. Alavi Nikje, M. Haghshenas and A. Bagheri Garmarudi, *Polymer-Plastics Technology and Engineering*, 2006, **45**, 4, 569. ©2006, Taylor & Francis [1]

Molecular weight distributions (M_WD) were determined using a 6A GPC instrument (Shimidzu). Water contents were measured using the Karl Fischer titration method. Viscosities were measured on a Haake™ viscometer (Thermo Scientific). pH was determined using a pH meter (Schott Geräte). All measurements were carried out under standard condition (25 °C, 1 atm).

For recycling on the laboratory scale, a mixture of DEG and DEA was employed as the reagent and solvent mixture, and sodium hydroxide (NaOH) was used as the catalyst, respectively. There have been various scientific sources reporting different conditions used to recycle PUF, but recycling of integral skin PUF has not been attempted. A reflux system containing a 1 L round-bottom flask equipped with a condenser, a thermometer, and an agitator was used. The glycolysis experiment was done at atmospheric pressure. A known mass of the solvent system was poured into the reactor and stirred at 700 rpm. The temperature of the heating mantel was set at 175 °C to dissolve the catalyst in the solvent and then increased to 200–205 °C and below the boiling point of DEG (244–245 °C). When the temperature of the mixture (DEG/DEA/NaOH) reached a specified value, the crushed integral skin PUF was poured in at a feeding rate of 4–5 g/ min. The glycolysis process was continued up to a reaction time of 5 h. Properties of the product were determined using instrumental and classical methods. Identical results had been obtained in a 1-m^3 reactor. The recovered polyol was formulated in combination with virgin polyol in steering-wheel and arm-rest production lines, and the prepared samples tested in an authorised polymer and chemistry laboratory. Obtained data revealed the performances of the recovered polyol for formulation of new final products.

3.1.3 Interpretations of obtained data

Having the recovered polyol as a liquid material without solid contaminants as well as unreacted foam wastes are key factors in chemical recycling. To ascertain the best conditions, in all experiments we tried to dissolve wastes under uncatalysed conditions. Results showed that dissolving integral skin PUF wastes in this condition was too difficult. Therefore, different catalysts were used in the solvent system to reduce the dissolution time. The solvent system contained DEG and DEA. Different catalysts were added to the solvent system and to find the best dissolution time for integral skin PUF. In each sample, 100 g of PUF was dissolved in 50 g solvent. We found that NaOH reduced the dissolution time the most. Reduction in dissolution time by various catalysts is shown in Table 3.3.

Table 3.3: Dissolution time of PU using different catalysts.

Catalyst	Dissolution time (min)
NaOH	10
KOH	14.5
KAc	57
NaAc	39
No catalyst	162

Reproduced with permission from M.M. Alavi Nikje, M. Haghshenas and A. Bagheri Garmarudi, *Polymer-Plastics Technology and Engineering*, 2006, **45**, 4, 569. ©2006, Taylor & Francis [1]

The solution was cooled too fast (immediately after dissolving the PUF in the solvent system). Then, GPC was used for the PUF, solvent system, virgin polyol, and fast-cooled solution. Chromatograms demonstrated conversion of PU to polyol by dissolution in the solvent system. One of the main goals in this research was to reduce energy and material losses, so various tests were done to find the best reaction temperature using NaOH as the catalyst: it was 200–205 °C. Results observed from various temperature conditions are shown in Table 3.4.

Table 3.4: Dissolution features at different temperatures.

Solution No.	PU (g)	Solvent (g)	Temperature (°C)	Time (min)	Result
1	10	10	150	20	Not dissolved
2	10	10	160	20	Not dissolved
3	10	10	170	20	Small amount dissolved
4	10	10	180	20	Dissolved in 17 min
5	10	10	190	20	Dissolved in 13 min
6	10	10	200	20	Dissolved in 10 min
7	10	10	210	20	Excellent dissolution with solvent evaporation
8	10	10	220	20	Solvent evaporated too fast

Reproduced with permission from M.M. Alavi Nikje, M. Haghshenas and A. Bagheri Garmarudi, *Polymer-Plastics Technology and Engineering*, 2006, **45**, 4, 569. ©2006, Taylor & Francis [1]

This process was continued for ≤5 h, then the product was collected from the process and left to cool down. As expected, the obtained product separated into two phases at room temperature (RT). The lower-phase was a viscous emulsion-dispersed, dark-brown liquid. The upper-phase was a light-brown liquid with low viscosity. GPC spectra of the virgin polyol demonstrated the M_WD to be 4,000–12,000 (66.87%) and a molecular weight (Mw) region was selected as a goal of the reaction end-point. The product was decanted and the upper-phase analysed by GPC to ensure that PU was converted and the Mw of the product was similar to that of virgin polyol (91.52%). The biggest advantage of this method is that both phases of the glycolysis product are useable in future foam formulations.

Main components of the lower-phase were glycols and amines. Therefore, the upper-phase was selected to be analysed and processed in further formulation of integral skin PUF. The lower-phase demonstrated acceptable results if used in formulation of rigid PUF.

After investigating the Mw of the product and confirming results, it is important to compare the recycled polyol with the virgin polyol and apply various quality-control tests:

a) *Viscosity*: Viscosity of the polyol is the most important physical parameter because it influences injection of the PUF. If the viscosity is too high, the polyol current will create a pressure that will damage the injection moulding machine. Hence, the viscosity of virgin and recycled polyols was determined at 25 °C. Viscosity of virgin polyol was 1,740 cps, whereas that of the recycled polyol was 1,870 cps. Accordingly, there was no appreciable difference between them in terms of viscosity.
b) *Specific gravity*: The specific gravity of the product and virgin polyol was measured using a pycnometer to observe differences during recycling. Specific gravity was 1.025 g/cm^3 for virgin samples and 1.069 g/cm^3 for recycled samples.
c) *Water content*: Water is a blowing agent that is very effective in PUF formulations. The water content of recycled polyol was measured (using the Karl Fischer titration method) and then compared with the water content of the virgin polyol. The water content of the recycled PUF was less than that of the virgin polyol. Hence, water needed to be added to reach water content of the virgin polyol (recycled polyol: 0.76%; virgin polyol: 1.16%).
d) *pH*: The pH of virgin polyol and recycling product as determined by a pH meter was 10.1 and 10.15, respectively.

3.1.4 New foam formulations using recycled polyols

To ascertain the maximum amount of recycled polyol applicable in PUF formulations, the recycled polyol was added to the foam formulation. As the percentage of recycled polyol was increased, more MDI had to be added. Results collected from this testing of foam formulations demonstrated a similarity between foams derived from virgin polyols and recycled polyols Tables 3.5 and 3.6.

Table 3.5: Material proportions in foam formulations for samples.

Sample	MDI (g)	Virgin polyol (g)	Recycled polyol (g)	CFC 11 (g)	Result
S1	40	100	–	4	Success
S2	40	70	30	4	Success
S3	60	70	30	7	Failed
S4	65	70	30	5	Success
S5	80	60	40	5	Success
S6	100	50	50	5	Success
S7	120	40	60	5	Success

CFC: Chlorofluorocarbon(s)
Reproduced with permission from M.M. Alavi Nikje, M. Haghshenas and A. Bagheri Garmarudi, *Polymer-Plastics Technology and Engineering*, 2006, **45**, 4, 569. ©2006, Taylor & Francis [1]

Table 3.6: Density and hardness of sample foams.

Sample	Density (g/cm^{-3}) free rise	Density (g/cm^{-3}) moulded	Hardness (shore A)
S1	188	209	85
S2	169.3	199.4	87 ± 1
S3	–	–	–
S4	142.6	203.3	93 ± 1
S5	124.4	208.3	95 ± 1
S6	120.1	200.7	95 ± 1
S7	119.6	206.1	94 ± 1

Reproduced with permission from M.M. Alavi Nikje, M. Haghshenas and A. Bagheri Garmarudi, *Polymer-Plastics Technology and Engineering*, 2006, **45**, 4, 569. ©2006, Taylor & Francis [1]

Another way to compare recycled polyols and virgin polyols is to examine their reactivity by measuring time data during foaming Table 3.7.

Table 3.7: Reactivity and time data for free rise integral skin PUF formulated using recycled polyol.

Sample	Cream time (s)	String time (s)	End of rise (s)
S1	50	88	107
S2	46	79	100
S3	Sample collapsed	–	–
S4	44	72	93
S5	47	79	96
S6	42	74	89
S7	43	77	93

Reproduced with permission from M.M. Alavi Nikje, M. Haghshenas and A. Bagheri Garmarudi, *Polymer-Plastics Technology and Engineering*, 2006, **45**, 4, 569. ©2006, Taylor & Francis [1]

Use of recycled polyols in PUF formulations did not result in appreciable differences (Figure 3.1), which demonstrated their applicability. All foaming processes were done in 20 ± 2 °C.

3.2 Trends in split-phase glycolysis processes

3.2.1 Introduction and background

PUs are one of the most versatile groups of polymer materials. PU types range from flexible/rigid foams and thermoplastic elastomers to adhesives, paints and varnishes.

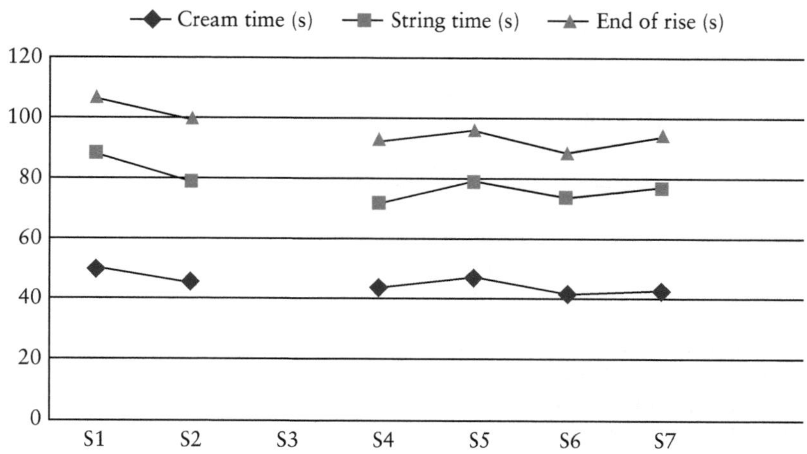

Figure 3.1: Comparison of the main reactivity factors of raw materials during the foaming process for virgin polyols and different amounts of recycled polyols. Reproduced with permission from M.M. Alavi Nikje, M. Haghshenas and A. Bagheri Garmarudi, *Polymer–Plastics Technology and Engineering*, 2006, **45**, 4, 569. ©2006, Taylor & Francis [1].

This variety is one of the reasons why the development of cost-efficient recycling processes is very difficult. PU manufacture is expensive compared with mass-produced plastic materials such as polyolefins, which is the reason for the development of PU recycling processes since the 1960s. In chemical recycling, glycolysis is one of the most important processes, which is used already on a commercial scale. The reaction products in glycolysis are polyols, and are used directly for the production of new PU.

In this research, the split-phase glycolysis reaction of cold-cure flexible PUF with DEG/DEA and NaOH was done to obtain a high-quality recycled polyol. Recovered polyols were used in new formulations for rigid and flexible foams [3].

3.2.2 General procedure of recycling

Waste cold-cure PUF received from automobile cushions was supplied from the Iranian Foam Company. It was formulated with the virgin polyol BAYNAT® PU 80IF04, which contained all the additives needed for the process and SUPRASEC® 5005 as an isocyanate (MDI). These raw materials were also used in subsequent formulations of flexible PUF after the glycolysis process because we wished to study the method in an industrial situation. The other virgin polyol (DALTOFOAM® TA 14,066) and MDI (SUPRASEC® 5005) used for the production of rigid foam and their technical data are listed in Tables 3.8–3.10.

The prefix 'virgin' is added to the main polyols to distinguish them from recycled polyol. The portion of each reagent in flexible and rigid PUF was:

Table 3.8: Technical data of virgin polyol used in cold-cure flexible PUF formulations.

BAYANT® PU 80IF04	
Appearance	**Semi-clear liquid**
Viscosity	1,740 cps at 25 °C
Specific gravity	1.025 g/cm^3 at 25 °C
Water content	1.16%
pH	10.10
Hydroxyl number	348 ± 20

Reproduced with permission from M.M. Alavi Nikje, M. Haghshenas and A. Bagheri Garmarudi, *Polymer-Plastics Technology and Engineering*, 2007, **46**, 3, 265. ©2007, Taylor & Francis [3]

Table 3.9: Technical data of virgin polyol used in rigid PUF formulations.

DALTOFOAM® TA 14,066	
Appearance	**Viscous yellow liquid**
Viscosity	5,263 cps at 25 °C
Specific gravity	1.06 g/cm^3 at 25 °C
Water content	2.3%
pH	9.45
Hydroxyl number	348 ± 20

Reproduced with permission from M.M. Alavi Nikje, M. Haghshenas and A. Bagheri Garmarudi, *Polymer-Plastics Technology and Engineering*, 2007, **46**, 3, 265. ©2007, Taylor & Francis [3]

Table 3.10: Technical data of MDI used in PUF formulations.

SUPRASEC® 5005	
Appearance	**Dark-brown liquid**
Viscosity	220 cps at 25 °C
Specific gravity	1.23 g/cm^{-3} at 25 °C
NCO value	30.9 wt% NCO groups (group wt: 42)
Average functionality	2.7
Flash point	233 °C
Fire point	245 °C

Reproduced with permission from M.M. Alavi Nikje, M. Haghshenas and A. Bagheri Garmarudi, *Polymer-Plastics Technology and Engineering*, 2007, **46**, 3, 265. ©2007, Taylor & Francis [3]

1. Cold-cure flexible PUF: 100 parts by weight (pbw) of BAYNAT® PU 80IF04 and 65 pbw of SUPRASEC® 5005.
2. Rigid PUF: 100 pbw of DALTOFOAM® TA 14,066 and 133 pbw of SUPRASEC® 5005.
3. We tried to omit use of CFC as a blowing agent because of its damaging effect on the ozone layer of the atmosphere. Hence, all samples and test formulations were free of CFC.

For the glycolysis process, a reactor system containing a three- necked round-bottomed flask (adapted with a reflux condenser) was equipped with a heater–agitator that could provide the chosen temperature and mixing. After carrying out the glycolysis process, virgin polyols and recycled polyols were characterised by:

a) The 6A GPC instrument (Shimadzu) for obtaining chromatograms.
   Conditions for GPC determination were: mobile-phase, THF; temperature, 40 °C; flow rate, 1 cm^3/min; injection volume, 100 µl; calibration, polystyrene standards (950, 2,200, 5,100, 11,600 Mw); detection, refractive index (RI).
b) The Karl Fischer titration method was used to determine water content.
c) A Haake™ viscometer (Thermo Scientific) was used for viscosity measurement.
d) A pH meter (Schott Geräte) was used for pH determination.
e) A Tensor 27 infrared (IR) spectrometer (Bruker) was used for infrared (IR) analyses.

A total weight of 200 g of the mixture of DEG and DEA was placed in a reaction flask, followed by mixing and addition of an appropriate amount of NaOH as the catalyst. The solvent system was heated up to 160 ± 5 °C as NaOH was dissolved in the mixture of DEG and DEA. Then, the temperature was increased up to 200 ± 5 °C. Waste cold-cure flexible PUF (400 g) was cut in small pieces (diameter, 1 cm) and added to the solvent system at a feeding speed of 15 g/min. Digestion occurred at an acceptable rate when the foam met the solvent. The glycolysis process was continued for 4 h under atmospheric pressure. Then, the products were separated into two (upper) clear-yellow, and (lower) light-brown phases after 24 h.

Phases were detached and a one-shot process applied for foam preparation. All ingredients were mixed and then MDI was added to allow curing. Polyol blends and all required additives were prepared in a 300-ml carton beaker by 2-min mixing followed by MDI addition. The pre-polymer was mixed for 10 s and then allowed to rise freely at RT, and was completely cured in the next 24 h.

3.2.3 Interpretations of obtained data

The digestion process of PU cold-cured foams was done with DEA/ diethylene amine as the glycolysing solvent system and NaOH as the catalyst. Digestion of PUF in a solvent system without a catalyst was extremely difficult. Therefore, different catalysts were employed in combination in the solvent system to reduce the digestion time. In each sample, 50 g of PUF was dissolved in 50 g of solvent and 0.5 g of catalyst. NaOH was found to have the greatest effect on reducing the dissolution time. Reduction in dissolution time by the various catalysts is shown in Table 3.11.

Table 3.11: Digestion time of waste PUF in solvent systems using different catalysts.

Catalyst	Dissolution time (min)
NaOH	9
KOH	15
KAc	59
NaAc	38
No catalyst	155

Reproduced with permission from M.M. Alavi Nikje, M. Haghshenasand A. Bagheri Garmarudi, *Polymer-Plastics Technology and Engineering*, 2007, **46**, 3, 265. ©2007, Taylor & Francis [3]

3.2.3.1 Identification and characterisation of recycled polyols

Reaction products were analysed by GPC. Comparison of the M_WD of virgin polyol with recycled-phases was carried out. Operating with a polymer:solvent ratio of 2:1 at 200 °C for 4 h elicited products separated into two upper and lower liquid-phases at a 65:35 ratio. Heating at 60 °C restored a single-phase. Secondary hydrogen bonds between the components seemed to be responsible for separation of a dense-phase with a relatively high viscosity at RT. The M_WD from GPC of virgin polyols and analog product-phases are shown in Table 3.12.

Industrial polyol standards (BAYNAT® PU 80IF04 and DALTOFOAM® TA 14,066) were employed to identify the liquid upper- and lower-phases, respectively. The upper liquid-phase was similar to the BAYNAT® PU 80IF04 polyol in the case of the cold-cure flexible polyol, and the lower-phase was similar to DALTOFOAM® TA 14,066 as the rigid PUF polyol. The upper-phase had a similar structure as the flexible PUF virgin polyol. The other part of the recycled polyols separated and migrated to the lower-phase (probably due to their higher specific gravities) to form a compound identical to the rigid PUF virgin polyol. The biggest advantage of this process is that both phases of the glycolysis product are useable in future PUF formulations.

It was important to investigate the similarity in chemical structure of virgin polyols and separated phases, so all samples were analysed by IR spectroscopy. IR spectra showed similarities in their functional groups. The Fourier-Transform

Table 3.12: M_WD of virgin polyols and analogue product phase derived from glycolysis reactions.

Time range (min)		Mw range		Abundance (%)	
From	To	From	To	BAYNAT	Upper phase
0	5.59	*	24,000	13.57	4.7
5.59	5.65	24,000	22,000	3.45	4.34
5.65	5.716	22,000	20,000	3.37	8.17
5.716	5.79	20,000	18,000	3.85	11.7
5.79	5.872	18,000	16,000	5.26	13.76
5.872	5.965	16,000	14,000	7.66	14.21
5.965	6.072	14,000	12,000	12.17	12.61
6.072	6.199	12,000	10,000	20.24	12.47
6.199	6.354	10,000	8,000	19.86	17.28
6.354	6.554	8,000	6,000	12.94	9.88
6.554	6.836	6,000	4,000	5.74	0.58
6.836	*	4,000	0	0.33	0.00
–	–	–	–	DALTOFOAM®	Upper phase
0	5.59	*	24,000	13.57	4.7
5.59	5.65	24,000	22,000	3.45	4.34
5.65	5.716	22,000	20,000	3.37	8.17
5.716	5.79	20,000	18,000	3.85	11.7
5.79	5.872	18,000	16,000	5.26	13.76
5.872	5.965	16,000	14,000	7.66	14.21
5.965	6.072	14,000	12,000	12.17	12.61
6.072	6.199	12,000	10,000	20.24	12.47
6.199	6.354	10,000	8,000	19.86	17.28
6.354	6.554	8,000	6,000	12.94	9.88
6.554	6.836	6,000	4,000	5.74	0.58
6.836	*	4,000	0	0.33	0.00

* Refers to the elution times longer than 6.836 min and Mw >2,400 and used for comparison of BAYNAT, DALTOFOAM® and upper phase components.

Reproduced with permission from M.M. Alavi Nikje, M. Haghshenas and A. Bagheri Garmarudi, *Polymer-Plastics Technology and Engineering*, 2007, **46**, 3, 265. ©2007, Taylor & Francis [3]

infrared (FTIR) spectra of a) BAYNAT® PU 80IF04, b) upper-phase, c) DALTOFOAM® TA 14,066, and d) lower-phase of recyclates are shown in Figure 3.2.

FTIR spectra of upper- and lower-phases revealed the similarity in chemical structures between recovered polyols, flexible polyols, and rigid virgin polyols. In the IR spectrum for BAYNAT® PU 80IF04 IR (a), intense stretching vibrations of the aliphatic ether group at 1,110 cm^{-1}, associated with hydroxyl end-groups and stretching vibrations of the hydroxyl group at 3,486 cm^{-1}, were noted. Absorption bands in the spectral region 2,870–2,971 cm^{-1} assigned to stretching vibrations of carbon–hydrogen bonds in aliphatic carbons, and in the spectral region 1,374 and 1,456 cm^{-1} assigned to bending vibrations of methylene groups in the polyol chain, were documented. In the FTIR spectrum of the upper-phase of recyclate b), stretching

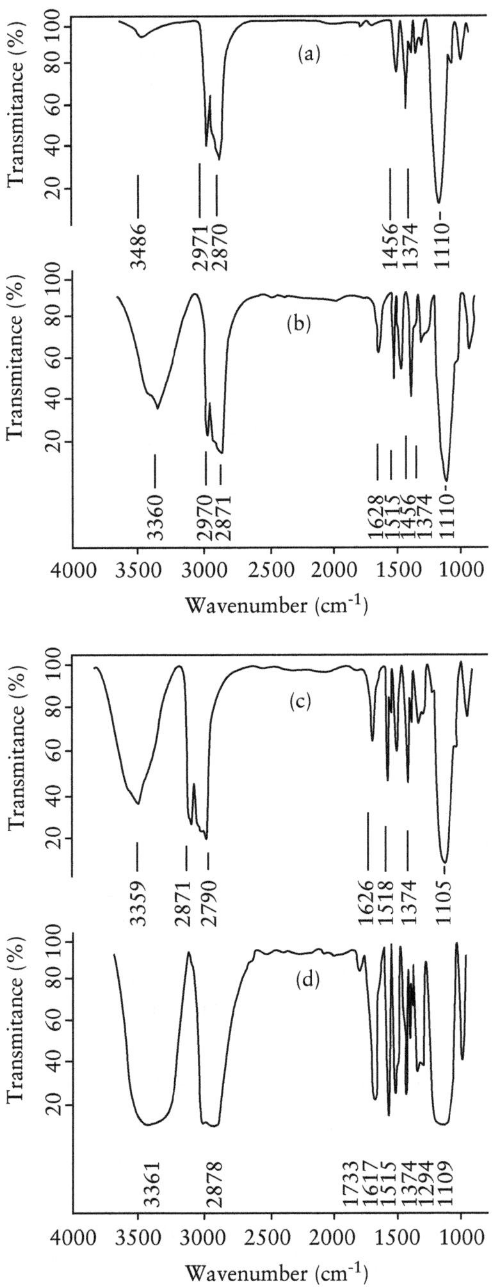

Figure 3.2: IR spectra of virgin polyols and recycled polyols. Reproduced with permission from M.M. Alavi Nikje, M. Haghshenas and A. Bagheri Garmarudi, *Polymer-Plastics Technology and Engineering*, 2007, **46**, 3, 265. ©2007, Taylor & Francis [3].

vibrations of C–O ether bonds at 1,110 cm^{-1}, hydroxyl groups at 2,871–2,971 cm^{-1} and methylene bending at 1,374–1,456 cm^{-1} were noted. FTIR spectrum of upper-phase b) is similar to the spectrum of BAYNAT® PU 80IF04 a), except for the increasing number of hydroxyl groups (absorption band at 3,360 cm^{-1}), N–H groups present in the DEA used as a catalyst, and other amine products produced in the glycolysis process.

These data suggested that the upper-phase obtained was polluted to a small extent by the bottom-phase because spectra demonstrated slight differences from the pure BAYNAT® PU 80IF04 polyol. Also, new absorption bands at 1,516 and 1,628 cm^{-1} were related to reaction byproducts that contaminated the upper-phase.

In the same comparison, in the FTIR spectrum of DALTOFOAM® TA 14,066 (c), intense stretching vibrations of the C–O ether and hydroxyl groups at 1,105 cm^{-1}, stretching vibrations of methylene groups at 2,871–2,970 cm^{-1}, and bending of methylene groups at 1,374, 1,467, 1,518 and 1,626 cm^{-1} were assigned additives in the formulated virgin polyol. Comparison of the FTIR spectra of lower-phase d) and DALTOFOAM® TA 14,066 c) revealed the similarity of the glycolysis-treated product to the virgin one. All bonds in the IR spectra of c) and d) were similar, except the splitting of the 3,359 cm^{-1} band in the lower-phase IR spectrum, which was due to amine products. Hence, the upper-phase could be used for formulations of flexible PUF at a defined percentage together with the virgin polyol.

On the other hand, the extracted polyol/ byproducts mixture in the lower-phase from the flexible PUF could be used for formulations of rigid PUF due to its similarity in structure to the virgin polyol (DALTOFOAM® TA 14,066). An additional weak absorption was detected in the FTIR spectrum of the recycled lower-phase d) at 1,733 cm^{-1} due to unreacted urea groups or oligomers.

After investigation of the Mw and IR analyses of on products, it was important to compare the recycled polyol components with virgin samples.

Comparative results of some quality-control tests are shown in Table 3.13. Polyol viscosity is the most physical parameter because it influences the PUF injection. If the viscosity is too high, the polyol current creates a pressure that can damage the injection moulding machine.

From Table 3.13 we can see that the viscosity of BAYNAT® PU 80IF04 is 1,740 cps, whereas the upper-phase of the recycled polyol has a viscosity of 1,870 cps. Also, the lower-phase has a viscosity that is ≈1,000 cps lower than that of DALTOFOAM® TA 14,066. Accordingly, an appreciable difference in viscosity was not observed.

Furthermore, the specific gravity of both phases of the product and virgin polyols was determined using a pycnometer. Specific gravity was 1.0250 and 1.0576 for virgin samples and 1.069 and 1.125 g cm^3 for recycled ones. Water as a blowing agent is another important factor in foam formulation, and was measured using the Karl Fischer titration method. Both product-phases had water content less than that of the virgin polyol, so water had to be added to reach the water content of the virgin polyol.

Table 3.13 revealed a relationship exists between the pH of BAYNAT® PU 80IF04 and the upper-phase of the recycled product. The pH (determined by a pH

Table 3.13: Result of quality-control tests on virgin and recycled polyols.

Parameter	Virgin polyol (BAYANT® PU 80IF04)	Recycle product (upper-phase)
Viscosity at 25 °C (cps)	1,740	1,870
Specific gravity at 25 °C	1.025	1.069
Water content (%)	1.16	0.76
pH	10.10	10.15
	Virgin polyol	**Recycled product**
Viscosity at 25 °C (cps)	5,263	4,120
Specific gravity at 25 °C	1.0576	1.125
Water content (%)	2.2858	0.856
pH	9.45	10.39

Reproduced with permission from M.M. Alavi Nikje, M. Haghshenas and A. Bagheri Garmarudi, *Polymer-Plastics Technology and Engineering*, 2007, **46**, 3, 265. ©2007, Taylor & Francis [3]

meter) was 10.10 for the virgin polyol and 10.15 for the recycled product. The difference between the pH of DALTOFOAM® TA 14,066 and lower-phase of the recyclate was because the lower-phase had a larger amount of basic compounds.

3.2.3.2 Reuse of recycled polyols in foam preparation

The formulation of a typical cold-cure flexible and rigid PUF used in our research is shown in Tables 3.14 and 3.15.

Table 3.14: Cold-cure flexible foam formulations: properties and reactivities.

Foam components (pbw)	S1	S2	S3	S4	S5	S6
Virgin polyol (DALTOFOAM® TA 14,066)	100	70	70	60	60	60
Recycled polyol (lower-phase)	0	30	30	40	40	40
MDI curing agent (SUPRASEC® 5005)	100	100	110	100	110	115
Free rise density (kg/m³)	44.31	53.03	45.59	76.19	57.12	49.35
Moulded density (kg/m³)	120.5	115.47	117.32	145.68	125.25	120.76
Cream time (s)	6	5	5	4	5	5
String time (s)	30	27	28	25	27	26
End of rise (s)	41	36	36	32	33	34
Tack-free time (s)	49	40	40	39	38	39

Reproduced with permission from M.M. Alavi Nikje, M. Haghshenas and A. Bagheri Garmarudi, *Polymer–Plastics Technology and Engineering*, 2007, **46**, 3, 265. ©2007, Taylor & Francis [3]

Table 3.15: Rigid foam formulations: properties and reactivities.

Foam components (pbw)	**S1**	**S2**	**S3**	**S4**	**S5**	**S6**	**S7**
Virgin polyol (DALTOFOAM® TA 14,066)	100	70	70	70	60	50	50
Recycled polyol (lower-phase)	0	30	30	30	40	50	50
MDI curing agent (SUPRASEC® 5005)	65	65	70	75	85	75	100
Free rise density (kg/m^3)	45.56	57.44	51.21	46.19	48.43	67.29	51.67
Moulded density (kg/m^3)	75.32	79.91	77.32	74.96	80.11	104.21	81.47
Cream time (s)	8	7	7	6	6	5	7
String time (s)	55	48	50	52	48	40	45
End of rise (s)	61	56	56	58	51	44	47
Tack free time (s)	79	70	72	66	65	56	58

Reproduced with permission from M.M. Alavi Nikje, M. Haghshenas and A. Bagheri Garmarudi, *Polymer–Plastics Technology and Engineering*, 2007, **46**, 3, 265. ©2007, Taylor & Francis [3]

Polyols obtained from the degradation of cold-cure foam during reaction injection moulding (RIM) were blended separately with the correlated virgin polyol at 30, 40, and 50% of total polyol. Levels >30% of recycled polyol could be tolerated by increasing MDI usage. Our detailed experiments on the recycled polyol show the presence of highly functionalised and low-Mw polyhydroxyl contained compounds which lead to the high hydroxyl values, as well as high functionality in comparison with virgin polyol.

Another way to compare recycled polyols and virgin polyols was to compare their reactivity using time data during foaming.

3.3 Extension of an ecofriendly method for chemical recycling of polyurethane foam wastes in 'split-phase'

3.3.1 Introduction and background

During recent years, the subject of PU recycling has been targeted, especially flexible PUF, due to its wide variety of applications, high volume of wastes, and as a way of solving environmental and economic problems. Cold-cure PUF constitute

29% of total production of foams. These foams are used widely in mattresses, furniture, and automotive seats. The glycolysis method for converting waste to raw materials has a special importance.

A chemical-recycling process for flexible PU based on split-phase glycolysis can be used to prepare high-quality flexible polyols that can be reused to make PUF with high recycled content. The split- phase glycolysis process has been targeted to recover flexible polyols of high quality and consistency to replace virgin polyols.

The chemistry of glycolysis involves breakdown of urethane bonds, in which the polymer chain is decomposed to a mixture of liquid oligomers, as shown in Equation 3.1 and 3.2:

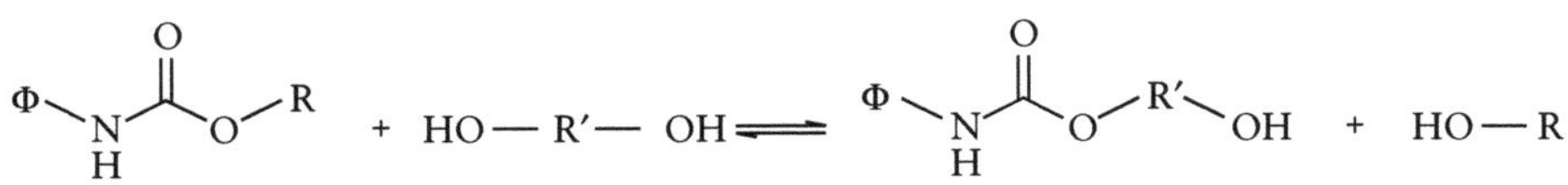

(3.1)

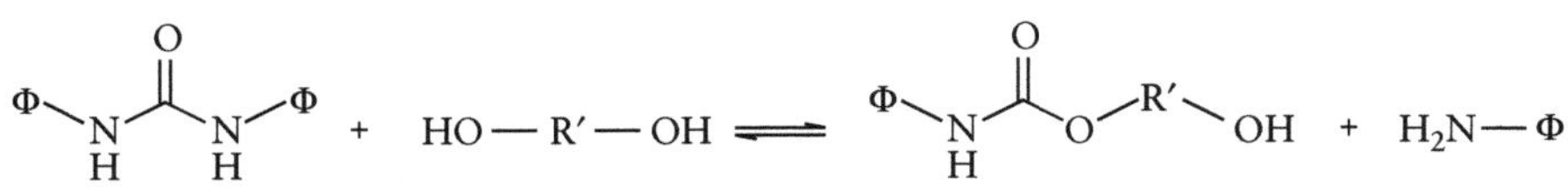

(3.2)

Urea bonds are also present in the PU structure, are formed in small quantities during PU synthesis, and result from the use of water as a foaming agent. Therefore, hydrolysis reactions can lead to the evolution of carbon dioxide (CO_2) and formation of amine groups, as shown in Equation 3.3:

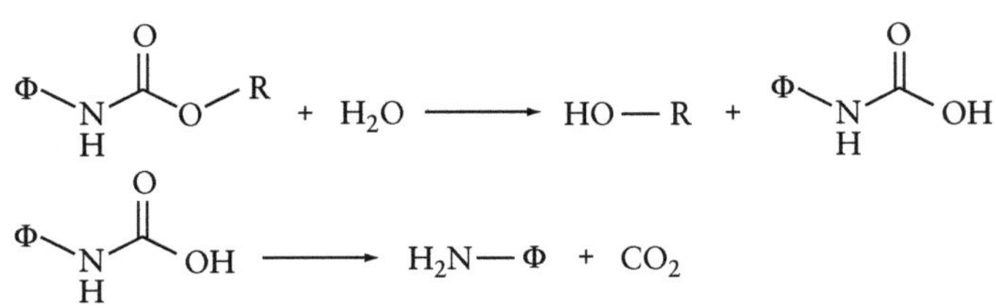

(3.3)

Transesterification reactions involving urea and urethane bonds and decarboxylation reactions lead to various byproducts, which can be increased due to pyrolysis reactions at the glycolysis temperature.

In this work, split-phase glycolysis of isocyanate-derived cold-cure foams, with glycerin as a new glycolysing agent, was studied. The process was simple and economically attractive among material- recycling methods. Glycerin was employed as a destroying solvent and NaOH as the catalyst. To study the ability of glycerin to be a glycolysing agent and for selection of a system to obtain high- quality recovered polyols, the effects of different reaction times were investigated [4].

3.3.2 General procedure of the recycling process

Industrial samples of flexible cold-cure PUF, which was derived from virgin polyol (Daltoflex® EC 20,240) were used. The formulation consisted of activated polyether polyols, aliphatic amines catalysts, water and a surfactant. The polyol was a colourless viscous liquid with a viscosity of 1.250 Pa.s at 20 °C, specific gravity of 1.035 g cm^3 at 20 °C, fire point of 240 °C [Mw ≈1,900, functionality of 2, and a hydroxyl number (OH#) of 59 mg potassium hydroxide (KOH)/g] and was in combination with SUPRASEC® 2027 MDI. The MDI was a dark-brown liquid with a viscosity of 0.220 Pa.s at 25 °C, specific gravity of 1.23 g/cm^3 at 25 °C, and NCO value of 30.9 wt% according to NCO group analyses (group weight: 42 g/mol, average functionality of 2.7, flash point of 233 °C, fire point of 245 °C). Daltoflex® EC 20,240 and SUPRASEC® 2027 are CFC-free systems (starting formulation was Daltoflex® EC 20,240 (100 pbw), SUPRASEC® 2027 (65 pbw); recommended mould temperature: 40–45 °C; demoulding time, 5 min; moulding density: 42–47 kg/m^3). The prefix 'virgin' is used to differentiate the main polyol (non- recycled polyol) from the recycled polyol. FTIR, ^{1}H-nuclear magnetic resonance (NMR) and ^{13}C-NMR spectroscopic methods were used for characterisation of recycled products and comparisons with the virgin polyol. FTIR spectroscopy was undertaken using a Tensor 27 spectrometer (Bruker). ^{1}H-NMR and ^{13}C-NMR spectroscopy were undertaken using a CRX 300 instrument (Bruker) with deuterated chloroform ($CDCl_3$) as a solvent. OH# were measured using a JNM- EX 90A NMR instrument (Jeol) according to the American Society for Testing and Materials (ASTM) D4273-94 method. The M_WD was measured using a GPC instrument (column type).

As a general procedure, a three-necked round-bottom flask was used. The flask was equipped with a mechanical stirrer and removable reflux condenser. The mixture of glycerin and NaOH was preheated until a homogeneous solution was obtained. PUF wastes segmented at arbitrary diameters (5–15 mm) were added gradually to the reaction mixture at 170–190 °C at 30 min. A transesterification reaction was carried out under atmospheric pressure and the reaction extended to complete dissolution of PUF. Reaction mixture was sampled at 30, 60, 90, 120, 150 and 180 min, transferred to a decanted funnel, and left to cool at RT. After several hours, the product separated into clear-yellow (upper) and dark-brown (lower) phases. Separated phases were analysed by classical and instrumental methods.

3.3.3 Interpretations of obtained data

To reduce the amine content of recycled products, glycerin was selected as a glycolysing agent and NaOH as a catalyst. Experiments were also followed in the absence of an amine decomposer. In a typical experiment, tri-phases were formed. The upper one constituted mainly the recovered polyol, and the bottom-phase transesterification by-products and excess glycerin. A third-phase was formed in a low amount (<10 pbw of the whole products of glycolysis) and was located between the upper-phase and lower-phase. It had a semi-solid state (gelatinous) and contained a mixture of worthless materials that were not characterised. Figure 3.3 shows the GPC chromatograms of sampled upper-phases at various reaction times.

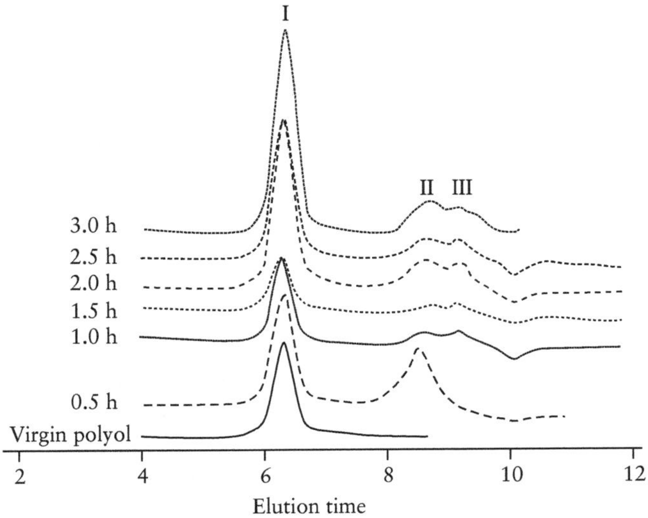

Figure 3.3: GPC chromatograms of virgin polyols and recycled polyols at different sampling times, I) polyol; II) transesterification byproducts; and III) glycerin. Reproduced with permission from M.M. Alavi Nikje and M. Nikrah, *Polymer Bulletin*, 2007, **58**, 2, 411. ©2007, Springer [4].

In Figure 3.3, peak I corresponds to the recovered starting polyol. The glycolysis process progressed well into the starting material (polyol) by a glycerin-based action that led to breakdown of PU chains into smaller fragments and, finally, complete degradation of PU functional groups. Transesterification byproducts were formed by substitution of the polyol by glycerin molecules (peak II). The intensity of this peak in the chromatogram corresponded to 0.5 h and, in comparison with the other peaks, was strong and wide. This finding demonstrated that, after a brief interval of perfect dissolution of foam, complete phase separation could not take place. Therefore, greater amounts of lower-phase products were in the upper-phase. Peak III corresponded to glycerin, which remained slightly in the upper-phase one. Figure 3.4a shows the GPC chromatograms of the industrial-starting polyether polyol used for PUF preparation and the reaction products obtained after complete decomposition of the PU.

According to the results obtained from GPC chromatograms, the starting polyol was recovered completely (it was the major product in the upper-phase with similar elution times and characteristics to those of the virgin polyol). On the other hand, peaks II and III in the chromatograms of the recovered polyol in the upper-phase

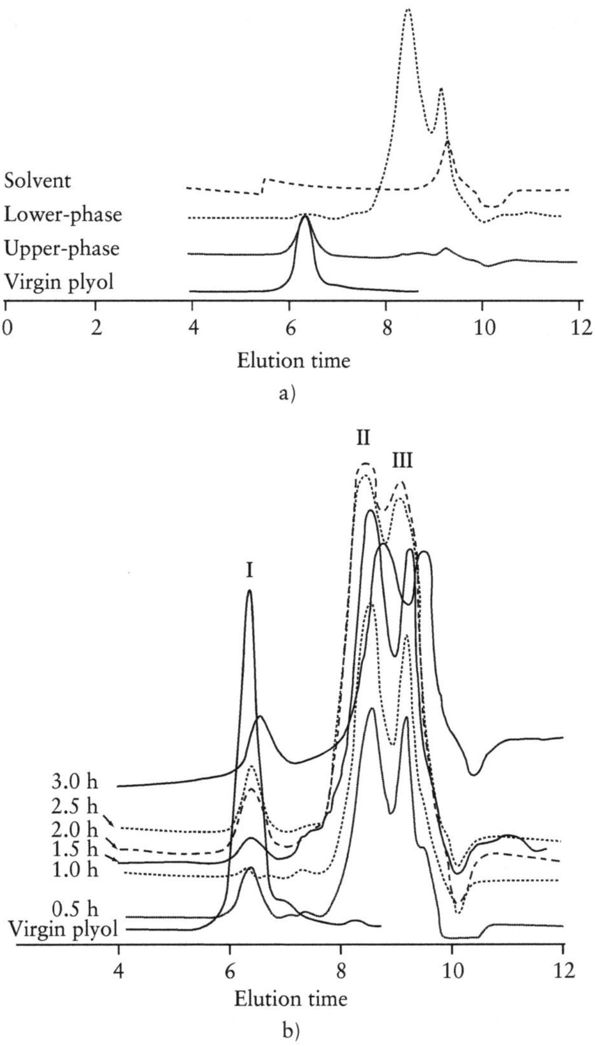

Figure 3.4: a) GPC chromatograms of 'split-phases' and their elution times compared with virgin polyol and b) GPC chromatograms of virgin polyol and lower-phase products at various reaction times. Reproduced with permission from M.M. Alavi Nikje and M. Nikrah, *Polymer Bulletin*, 2007, **58**, 2, 411. ©2007, Springer [4].

demonstrate product pollution (with the bottom liquid layer having a low-Mw because of very low miscibility). Figure 3.4b shows the chromatograms obtained for the bottom liquid-phase and virgin polyol.

The bottom-phase was enriched by highly polar products (mainly excess glycerin and aromatic byproducts derived from the starting isocyanates). A small quantity of the lower-phase was dissolved in the upper-phase, which reduced its purity.

The chromatogram of the bottom-phase obtained with glycerin demonstrated the characteristic peak of polyol (I). Figure 3.5 shows the FTIR spectrum of the virgin polyol used in the formation of the PUF a) as well as the upper-phase after 1 h b) and 3 h c), from perfect foam dissolution.

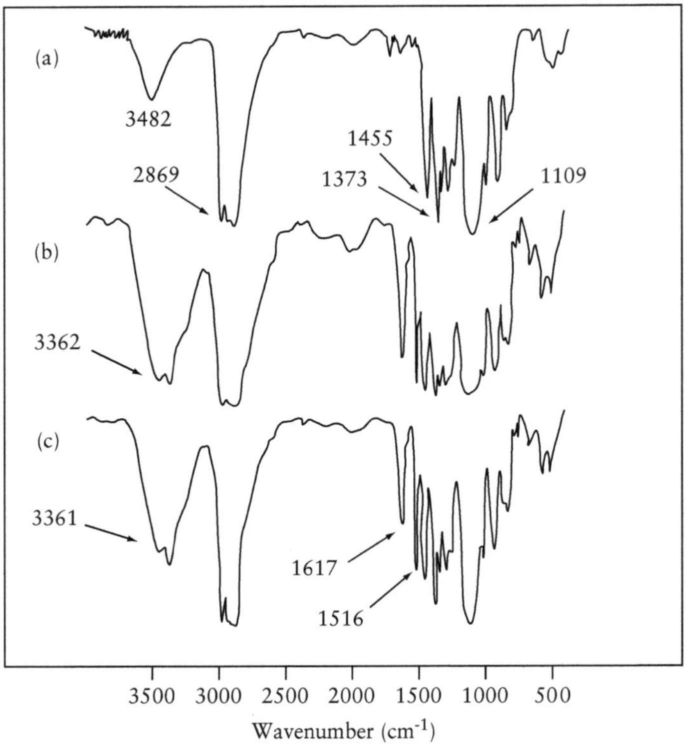

Figure 3.5: Comparisons of IR spectra for: virgin polyol a), recycled polyol after 1 h b) and 3 h c), from perfect foam dissolution. Reproduced with permission from M.M. Alavi Nikje and M. Nikrah, *Polymer Bulletin*, 2007, **58**, 2, 411. ©2007, Springer [4].

The peak in the spectral region of 1,109 cm^{-1} corresponded to the stretching vibrations of the aliphatic ether groups, together with the bond formed by CO groups associated with hydroxyl end-groups as overlapping and stretching vibrations of these hydroxyl groups at 3,482 cm^{-1}. The latter band was broader in b) and c) in comparison with virgin one a) due to primary amine contaminants (Equations 3.2 and 3.3) in the upper-phase.

Absorption bands in the spectral region 2,869–3,000 cm^{-1} were related to the stretching vibrations of CH bonds in aliphatic carbons, and those at 1,455 and 1,373 cm^{-1} were characteristic of bending vibrations of methylene and methyl groups in the polyol chain. Comparison of IR spectra of virgin and recyclated polyols demonstrated their structural similarity. Only three differences were noted between recovered and virgin polyols. First, absorption bands at >3,000 cm^{-1} in b) and c) were widened by NH groups in the upper- phase as contaminants [absorption bands at 3,362 and 3,361 cm^{-1} in b) and c), respectively]. Second, new absorption bands at 1,617 and 1,516 cm^{-1}, in b) and c) spectra were related to the bending vibrations in the amine group slightly dissolved in the upper-phase. Third, glycerin was dissolved in the upper-phase in small amounts but, because all stretching vibrations of glycerin bonds (C-O and OH bonds) were in the same spectral regions, C-O and OH bonds of the starting polyol appeared in the FTIR spectrum.

Hence, no observable changes in absorption bands were noted except for an increase in the quantity of hydroxyl groups that lead to increases in the intensity of absorption bands at 3,362 and 3,361 cm^{-1} in b) and c), respectively.

Bonds assigned to the stretching vibrations of the carbonyl groups in urethane linkages appeared at 1,690–1,749 cm^{-1}. Absorption bands were not found in these regions in b) and c) spectra, so completion of urethane-bond breaking occurred 1 h after perfect dissolution of the PUF.

^{1}H-NMR and ^{13}C-NMR spectra of recycled and virgin polyols were investigated. Figure 3.6 shows the ^{1}H-NMR a) and ^{13}C-NMR b) spectra of virgin polyol. Most of the bond assignments in ^{13}C-NMR spectra have been indicated.

In the ^{1}H-NMR spectrum, the peak in the region of 1.1 parts per million (ppm) was related to hydrogen atoms in methyl groups. Peaks at 3.0–4.0 ppm corresponded to hydrogen atoms and carbon atoms connected directly to oxygen atoms. Comparison of the ^{1}H-NMR spectra of virgin polyol and recycled product in the upper- phase (Figures 3.7) suggested that the chemical structures of these compounds were very similar, except that the peaks at 6.6–7.1 ppm and 6.9–7.2 ppm (Figures 3.7b and **c**, respectively) were related to aromatic-ring hydrogen atoms derived from aromatic diamines that remained partially in the recovered polyol.

The only difference between the ^{13}C-NMR spectra of recovered and virgin polyols (Figure 3.8) was signals in the low peak area (110–140 ppm) for recycled polyols originating from the starting isocyanate as a contaminant.

Figure 3.9 demonstrates the IR spectrum of the lower-phase that contained a mixture of carbamate, urea, decarboxylated products and excess glycerin. Absorption bands in the spectral regions 3,418, 1,614 and 1,516 cm^{-1} at high intensity confirm the presence of urea, carbamate and similar functional groups in the bottom-phase in higher concentrations than those in the upper-phase. Comparison of a) and b) spectra showed no significant reactions after 1 h to 3 h from perfect dissolution of PU, so the obtained products in specified intervals were quite equal. Obtained results from GPC, NMR and IR confirmed complete degradation of urethane chains 1 h after complete dissolution of the foam. However, to reassure ourselves that the reaction had been complete, the reaction went on until 3 h. Obtained results showed that, after 1 h, no noticeable change was observed in glycolysis products and/or in the upper-phase: lower-phase ratio. This finding showed the susceptibility of glycerin together with

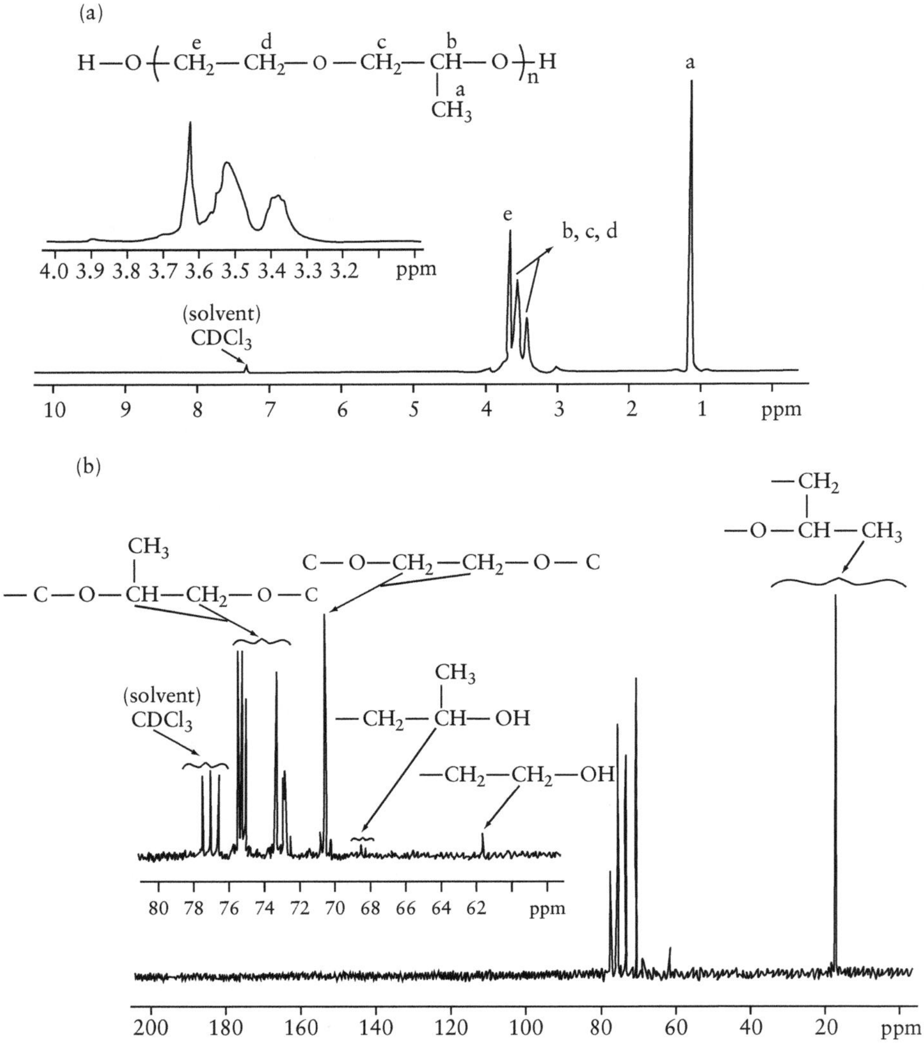

Figure 3.6: ^{1}H-NMR a) and ^{13}C-NMR b) spectra of virgin polyol. Reproduced with permission from M.M. Alavi Nikje and M. Nikrah, *Polymer Bulletin*, 2007, **58**, 2, 411. ©2007, Springer [4].

NaOH as a strong destroying agent in degradation of urethane chains in a short time after complete dissolution of the foam. Another method for comparison of recycled and non-recycled polyols was OH# determination. According to ASTM D4273-94 by ^{13}C-NMR, polyol (0.6 g) was dissolved in $CDCl_3$ (0.4 g) as a solvent (sample/ solvent ratio, 3:2), mixed thoroughly, and added to the NMR tube.

A sufficient number of repetitive pulses were accumulated using a NMR 90-MHz spectrometer until the peaks of primary and secondary hydroxyl carbons could be

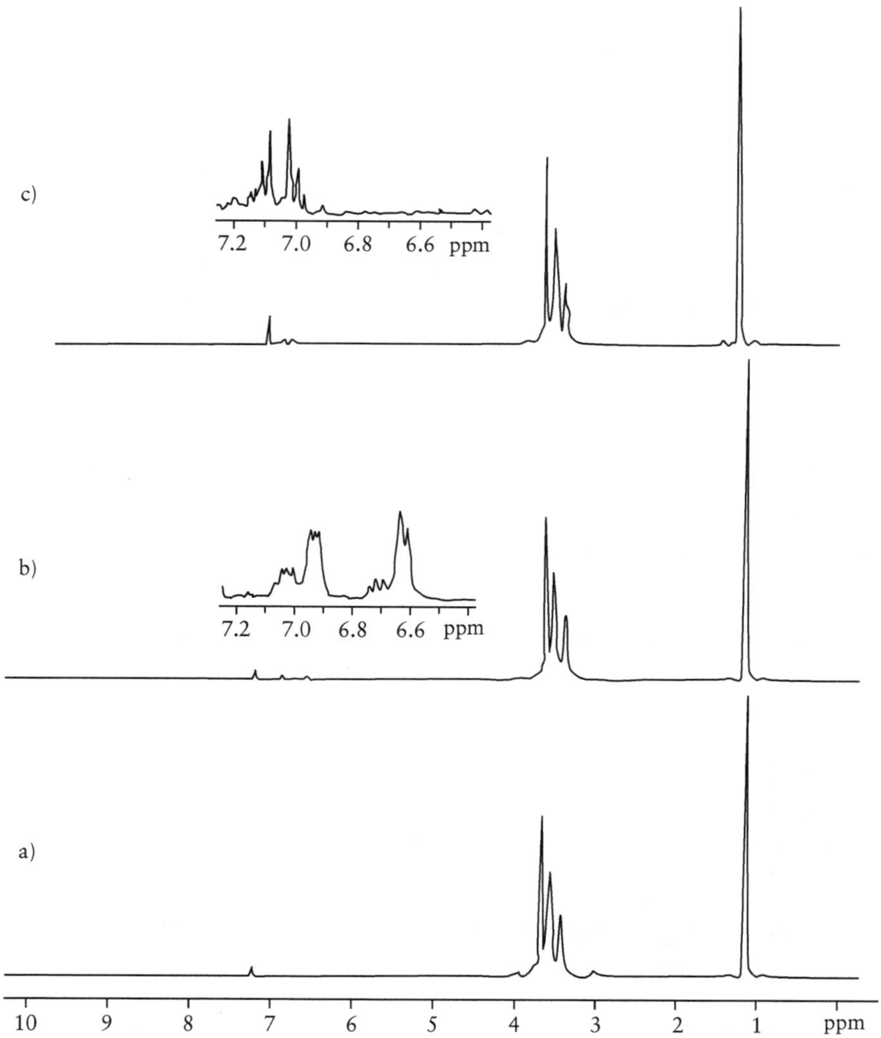

Figure 3.7: ^{1}H-NMR spectra of virgin a) and recycled polyols after 1 h b) and 3 h c), from perfect foam dissolution. Reproduced with permission from M.M. Alavi Nikje and M. Nikrah, *Polymer Bulletin*, 2007, **58**, 2, 411. ©2007, Springer [4].

determined accurately by the spectrometer's integration system. The peaks appeared in the region 60–68 ppm in the ^{13}C-NMR spectrum (which were assigned previously in Figure 3.6b). As shown in Figure 3.10, ^{13}C-NMR spectra of virgin a) and recycled polyols after 1 h b) and 3 h c), from perfect foam dissolution were expanded, amplified and integrated for determination of OH#. Peaks of primary and secondary hydroxyl carbons corresponded to the glycerin remaining in the recycled polyol, and so appeared in the region 58–68 ppm (the latter region in Figures 3.10b and c is more crowded because

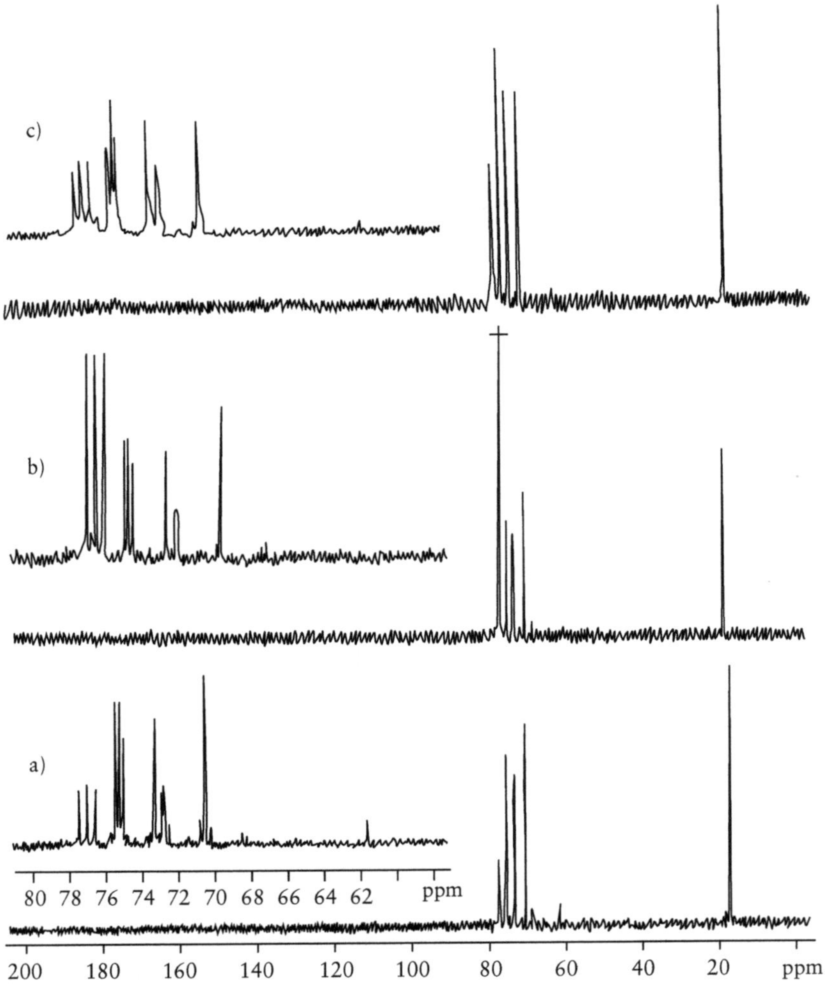

Figure 3.8: Comparison of ^{13}C-NMR spectra of virgin a) and recycled polyol after 1 h b) and 3 h c), from perfect foam dissolution. Reproduced with permission from M.M. Alavi Nikje and M. Nikrah, *Polymer Bulletin*, 2007, **58**, 2, 411. ©2007, Springer [4].

presumably some of the peaks overlapped). OH% for any sample was calculated according to Equations 3.4 and 3.5:

$$\text{OH\%} = \frac{\text{Peak areas of primary OH groups} + \text{peak areas of seconday OH groups}}{\text{All peak areas in spectrum}} \tag{3.4}$$

$$\text{OH\#} = \text{OH\%} \times 33 \tag{3.5}$$

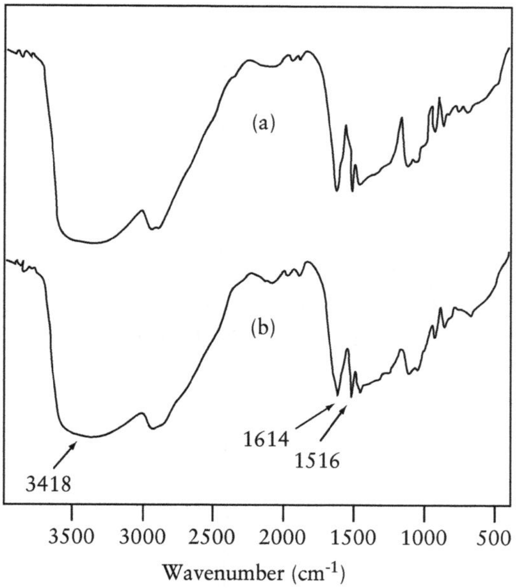

Figure 3.9: IR spectra of lower-phases: after 1 h (a) and 3 h (b) from perfect foam dissolution. Reproduced with permission from M.M. Alavi Nikje and M. Nikrah, *Polymer Bulletin*, 2007, **58**, 2, 411. ©2007, Springer [4].

The OH# of recycled polyols was higher due to partial solubility of glycerin in the upper-phase in comparison with virgin polyol (59 mg KOH.g^{-1}). Hydroxyl values for the recycled polyols 1 h and 3 h after complete dissolution of the foam were determined: 99 and 73 mg KOH.g^{-1}, respectively. On the other hand, when the reaction progressed, the concentration of glycerin in the upper-phase decreased, which caused a decreasing in hydroxyl values. Therefore, the obtained polyol after 3 h had a lower hydroxyl value than the virgin polyol after 1 h.

Another observation was CO_2 evolution from the reaction mixture. This process was because of mass losses at the reaction end-point and gas bubbling in cold NaOH solution.

Viscosity is the most important physical parameter of a polyol and influences PUF injection. Viscosities of virgin polyols and reaction mixtures at different reaction times are shown in Table 3.16. We can see that the viscosity of Daltoflex® EC 20,240 is 1,250 cps, whereas the reaction mixture has a maximum viscosity of 9,600 cps at 30-min reaction time. On the other hand, viscosity decreased upon increase in reaction time due to successful scissions of PU chains.

Obtained results from viscosity studies confirmed data from GPC, NMR and IR. They showed complete degradation of urethane chains 1 h after perfect dissolution of the foam. On the other hand, there was no considerable difference in viscosity after 1 h.

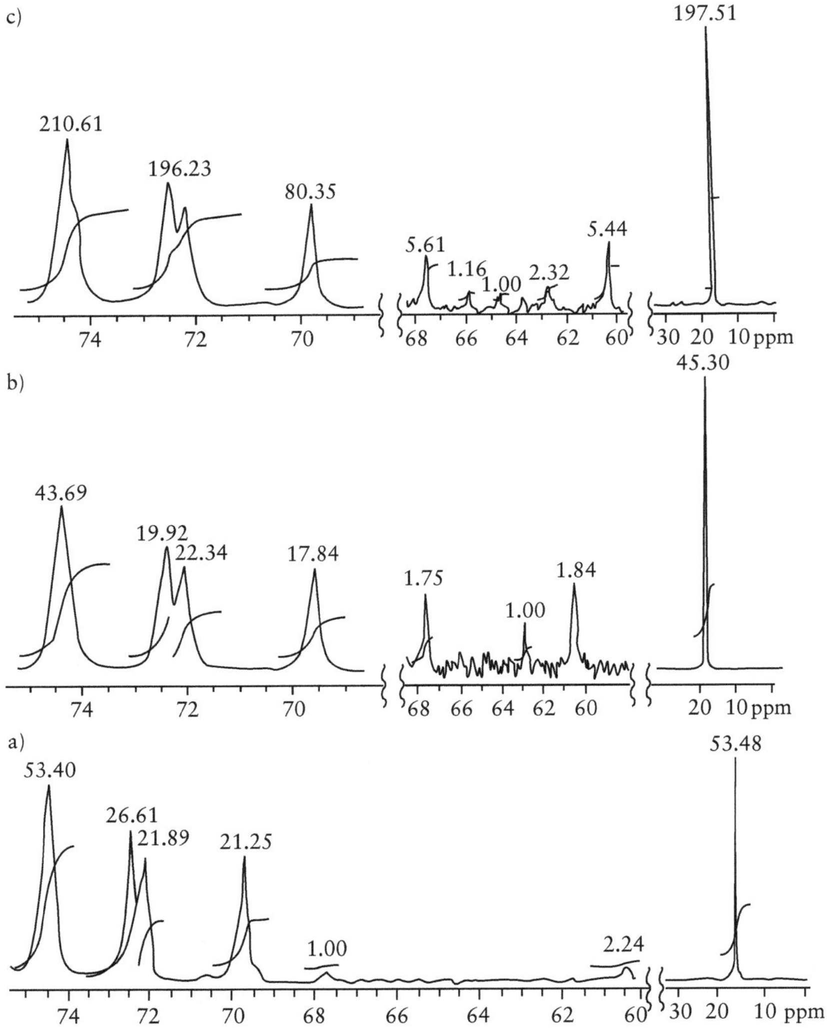

Figure 3.10: ^{13}C-NMR spectra of virgin a) and recycled polyols after 1 h b) and 3 h c), from perfect foam dissolution for determination of OH number. Reproduced with permission from M.M. Alavi Nikje and M. Nikrah, *Polymer Bulletin*, 2007, **58**, 2, 411. ©2007, Springer [4].

3.4 Split-phase glycolysis of polyurethane flexible foam wastes under molecular-weight irradiation

3.4.1 Introduction and background

Numerous applications in material-processing methods have been introduced that have resulted in shorter reaction times and greater convenience. Microwave (MW)

Table 3.16: Changes in viscosity of the reaction mixture versus time.

Reaction time (h)	**Reaction mixture viscosity (cps)**
0.5	9,600
1.0	3,560
1.5	3,412
2.0	3,425
2.5	3,400
3.0	3,450
Virgin polyol	1,250

Reproduced with permission from M.M. Alavi Nikje and M. Nikrah, *Polymer Bulletin*, 2007, **58**, 2, 411. ©2007, Springer [4]

heating offers reduction in electrical consumption and increases in reaction rates, compared with thermal heating. The reaction rate is increased in MW conditions due to a higher rate of heat transfer and formation of 'hotspots' in comparison with conventional heating methods. Glycerin has been accepted as a 'green' solvent, so we explored chemical recycling using this reagent in combination with a 'clean' energy resource: MW energy. The main merit of this method comes from undertaking the reaction under atmospheric pressure [5].

3.4.2 General procedure of the recycling process

The materials used in our study were selected from flexible cold-cure PUF formulated by Daltoflex® EC 20,240 polyol and SUPRASEC® 2027 MDI. A MicroSYNTH 'NP' Ethos 1600 MW oven (Milestone) was used as the energy source. FTIR, ^{1}H-NMR and ^{13}C-NMR spectroscopic methods were used for characterisation of recycled products and comparisons with the virgin polyol. ^{1}H-NMR and ^{13}C-NMR spectra were recorded using a CRX 300 instrument (Bruker) and acetone-d_6 was the solvent.

In an experiment, glycerin and catalyst in specified portions were placed in a reaction flask and irradiated in the MW oven until a homogeneous solution was obtained. After cooling at RT, ground waste PUF (particle diameter of 5–10 mm in the equal weight of solvent and catalyst) was added to the reaction mixture, and exposed to irradiation. To examine the effect of the reaction temperature, the glycolysis process was undertaken at 160, 180, 200 and 220 °C in the MW-irradiation programmes detailed in Table 3.17.

Complete disappearance of PUF particles as the final dissolution time was the aim of the study. In some reactions, additional times of 10–30 s after complete dissolution of the foam was needed to achieve complete breakdown of urethane linkages.

Table 3.17: Applied MW irradiation programmes in glycolysis at different reaction temperatures.

Fixed temperature (°C)	Step	Time (min)	Temperature (°C)	Maximum power (watt)
160	1	2	Ramp to 160	Up to 700
	2	2	Hold at 160	Up to 400
180	1	1.5	Ramp to 180	Up to 800
	2	2	Hold at 180	Up to 800
200	1	1.5	Ramp to 200	Up to 900
	2	2	Hold at 200	Up to 500
220	1	1.5	Ramp to 220	Up to 1,000
	2	2	Hold at 220	Up to 600

Reproduced with permission from M.M. Alavi Nikje, M. Nikrah and M. Haghshenas, *Polymer Bulletin*, 2007, **59**, 1, 91. ©2007, Springer [5]

After completion of the reaction, the mixture was removed from the MW oven and filtered to isolate solid contaminants and unreacted foam. Filtered liquid was allowed to separate into two phases. The upper-phase was the recovered polyol and the lower one consisted of transesterification byproducts and unreacted glycerin. The upper-phase was decanted carefully, centrifuged for 20 min, and characterised by spectroscopic methods.

3.4.3 Interpretations of obtained data

The minimum energy consumption for reaction completion is the main criterion in organic chemistry. Hence, in all of our experiments, the effect of MW irradiation on the glycolysis reactions, different reaction times, various temperatures and catalysts were investigated. In this work, glycerin together with KOH or NaOH was selected as the destroying agent. To find the best dissolution time, all reactions were repeated thrice. The obtained average reaction times are listed in Table 3.18.

Table 3.18: Foam dissolution times (s) at various temperatures using different catalysts.

Temperature (°C)	KOH	NaOH
160	110	93
180	92	81
200	78	79
220	67	65

Reproduced with permission from M.M. Alavi Nikje, M. Nikrah and M. Haghshenas, *Polymer Bulletin*, 2007, **59**, 1, 91. ©2007, Springer [5]

The transesterification reaction was carried out using a glycerin/ NaOH or KOH system, and led to chain fracture of PU linkages according to Equation 3.6:

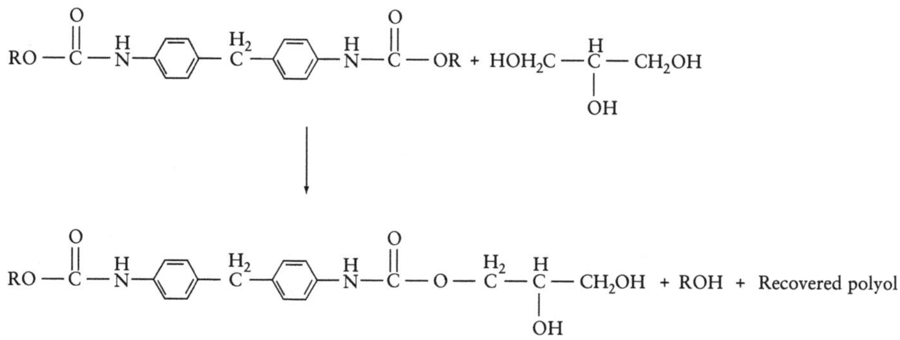

(3.6)

After completion of the reaction, split-phases were obtained. Obtained data from characterisation and identification of the reaction products revealed that the lower-phase contained a broad variety of byproducts obtained from the transesterification reaction of urea, urethane and allophanate functional groups as well as decarboxylation byproducts. Hence, using water as the foaming agent in the production of PUF leads to a reaction with the NCO group to form the unstable intermediate carbamic acid. This process results in elimination of CO_2 and formation of a primary amine group that reacts with the NCO functional group to form urea bonds, as shown in Equations 3.7 and 3.8:

$$Ar-NCO + H_2O \longrightarrow \underbrace{\left[Ar-\overset{H}{N}-\overset{O}{\overset{\|}{C}}-OH\right]}_{\text{Carbamic acid}} \longrightarrow Ar-NH_2 + CO_2 \tag{3.7}$$

The primary amine group reacts immediately with an additional isocyanate group, followed by formation of urea bonds:

$$Ar-NCO + Ar'-NH_2 \longrightarrow \underbrace{Ar-\overset{H}{N}-\overset{O}{\overset{\|}{C}}-\overset{H}{N}-Ar'}_{\text{Disubstituted urea}} \tag{3.8}$$

The upper-phase contained recovered polyol and was characterised and identified by spectroscopic methods, and data compared with results for the virgin polyol. For comparison, Figure 3.11 demonstrates the FTIR spectrum of the Daltoflex® EC 20,240 virgin polyol. As shown in Figure 3.11, an absorption band at 1,115 cm^{-1} is due to the stretching vibrations of the aliphatic ether groups together with the signal manufactured by CO groups associated with hydroxyl end-groups, and the overlapping and stretching vibrations of these hydroxyl groups is at 3,482 cm^{-1}.

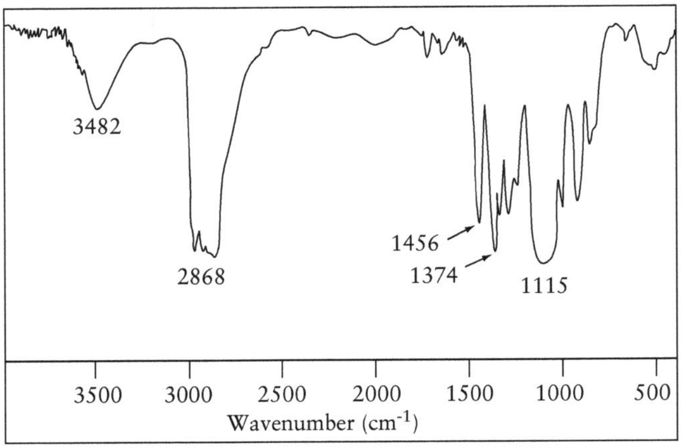

Figure 3.11: FTIR spectrum of the Daltoflex® EC 20,240 polyol. Reproduced with permission from M.M. Alavi Nikje, M. Nikrah and M. Haghshenas, *Polymer Bulletin*, 2007, **59**, 1, 91. ©2007, Springer [5].

Absorption bands in the 3,000–2,868 cm^{-1} region are related to the stretching vibrations of CH bonds. Bands at 1,456 and 1,374 cm^{-1} are related to the bending vibrations of methylene and methyl groups in the polyol chain.

FTIR spectra of recovered polyols at different reaction temperatures and in the presence of KOH or NaOH catalysts were similar to those of the virgin polyol (Figures 3.12 and 3.13, respectively).

Comparison of the FTIR spectra in Figures 3.12 and 3.13 revealed only a few slight differences between recovered and virgin polyols. Main differences arose from the broader absorption bands at >3,000 cm^{-1} for the recovered polyol, which were related to the stretching vibrations of NH_2 functional groups of primary amines as contaminants in the recovered polyol. 4,4′-Methylenediphenyl diamine was one of the reaction byproducts obtained during the glycolysis process due to transesterification of urea groups present in urethane chains or the hydrolysis reaction. Trapped water molecules in PU wastes are responsible for this reaction, as shown in Equations 3.9 and 3.10.

$$\mathrm{Ar-NH-C(=O)-NH-Ar' + HOH_2C-CH(CH_2OH)-CH_2OH \longrightarrow Ar'-NH_2 + Ar-NH-C(=O)-O-CH_2-CH(CH_2OH)-CH_2OH} \quad (3.9)$$

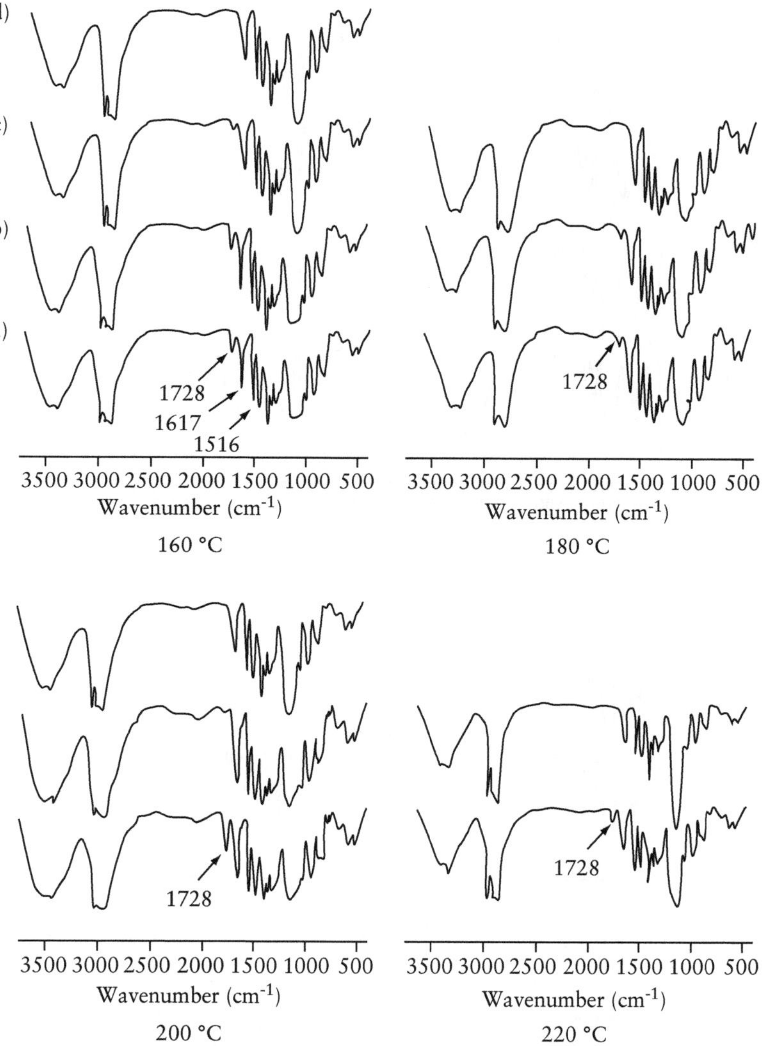

Figure 3.12: FTIR spectra of recovered polyols at different temperatures in the presence of KOH as a catalyst. Extended reaction times for glycolysis completion are: a) 0; b) 10; c) 20; and d) 30 s after perfect foam dissolution. Reproduced with permission from M.M. Alavi Nikje, M. Nikrah and M. Haghshenas, *Polymer Bulletin*, 2007, **59**, 1, 91. ©2007, Springer [5].

Therefore, new absorption bands at 1,617 and 1,516 cm^{-1} corresponded to the bending vibrations of amine groups derived from 4,4′-methylenediphenyl diamine, which was slightly dissolved in the upper-phase. Although small portion of glycerin was dissolved in the upper-phase, no appreciable changes in absorption bands appeared except for signal widening at >3,000 cm^{-1}. All stretching vibrations of

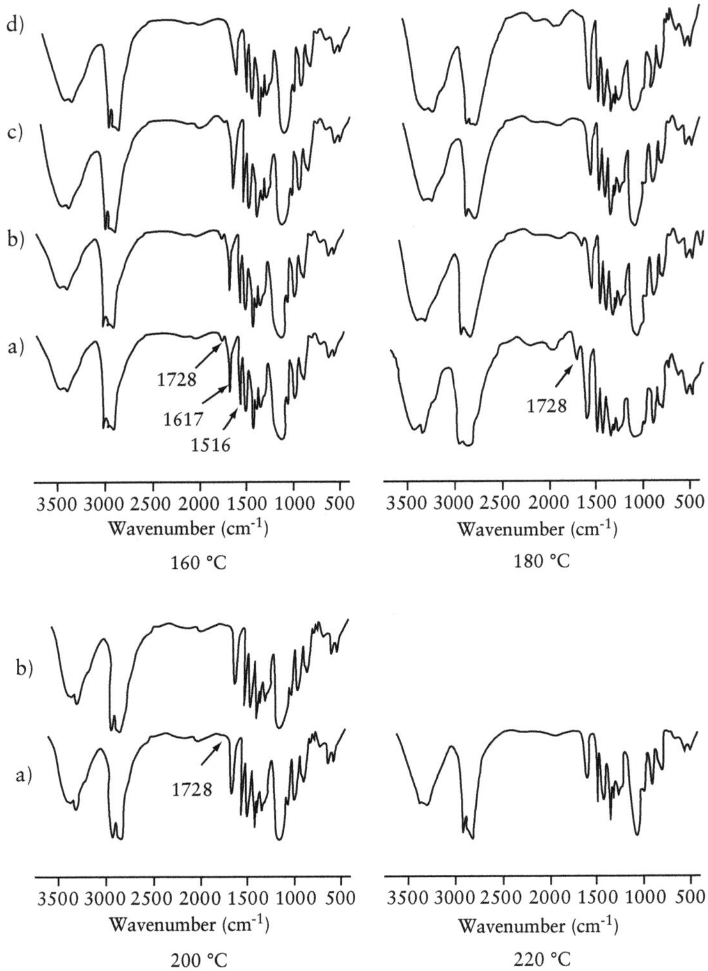

Figure 3.13: FTIR spectra of recovered polyols at different temperatures in the presence of NaOH as a catalyst. Extended reaction times for complete glycolysis are: a) 0; b) 10; c) 20; and d) 30 s after perfect foam dissolution. Reproduced with permission from M.M. Alavi Nikje, M. Nikrah and M. Haghshenas, *Polymer Bulletin*, 2007, **59**, 1, 91. ©2007, Springer [5].

glycerin bonds (C-O and OH bonds) were in the same spectral regions of C-O and OH bonds of the virgin polyol.

Bands assigned to the stretching vibrations of the carbonyl groups in urethane linkages were found at 1,749–1,690 cm^{-1}. Therefore, the stretching vibrations in the spectral region 1,728 cm^{-1} shown in Figures 3.12 and 3.13 were related to urethane groups, the intensity of which decreased with increasing reaction time. The glycolysis process was completed with disappearance of stretching vibrations of urethane groups at 1,728 cm^{-1}. One of the most important findings in these reactions was the dependence of disappearance times of urethane bands to the catalyst type and reaction temperature. These results were concluded from Figures 3.12 and 3.13. In Figure 3.12, in the presence of KOH, the reaction time decreased from 30 s to 10 s when the temperature rose from 160 °C to 220 °C. Use of NaOH as a catalyst at higher temperature (220 °C) represented perfect destruction of polymer chains immediately after perfect dissolution of foams (Figure 3.13).

Figure 3.14 shows the FTIR spectra of the lower-phase sampled at various reaction times. Similar to the upper-phase, absorption bands at 1,614 and 1,515 cm^{-1}, as well as two weak signals at 512 and 572 cm^{-1}, confirmed formation of amine compounds which had concentrated in the reaction mixture with increasing time. NMR spectroscopy was employed for investigation of the chemical structure of the recovered polyol and compared with that of the virgin polyol.

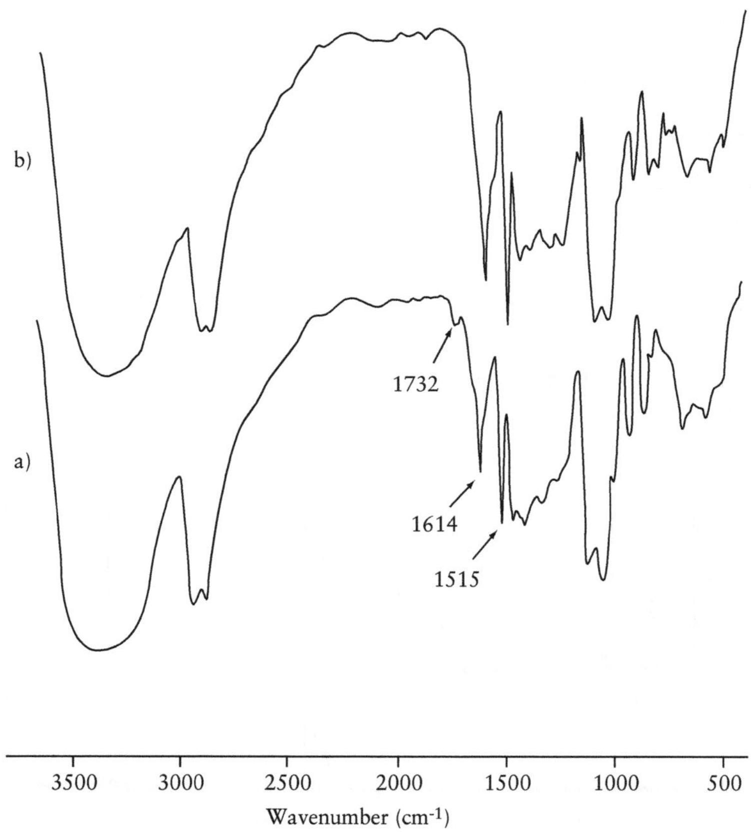

Figure 3.14: Comparison of FTIR spectra of obtained products in the lower-phase in the presence of NaOH at 160 °C and at different reaction times: a) 0 and b) 30 s after perfect foam dissolution. Reproduced with permission from M.M. Alavi Nikje, M. Nikrah and M. Haghshenas, *Polymer Bulletin*, 2007, **59**, 1, 91. ©2007, Springer [5].

Figure 3.15 shows the ^{1}H-NMR and ^{13}C-NMR spectra of the virgin polyol. In the ^{1}H-NMR spectrum, the peak at 1.1 ppm is related to methyl hydrogen atoms, and the peaks at 3.0–4.0 ppm correspond to hydrogen atoms directly connected to C-O groups. The quintet peak in 2.05 ppm is related to deuterated acetone. ^{1}H-NMR and ^{13}C-NMR spectra of recovered polyols at lowest and highest reaction temperatures (160 and 220 °C) at the endpoint of the reaction (complete breakdown of urethane linkages) are shown in Figures 3.16 and 3.19, respectively.

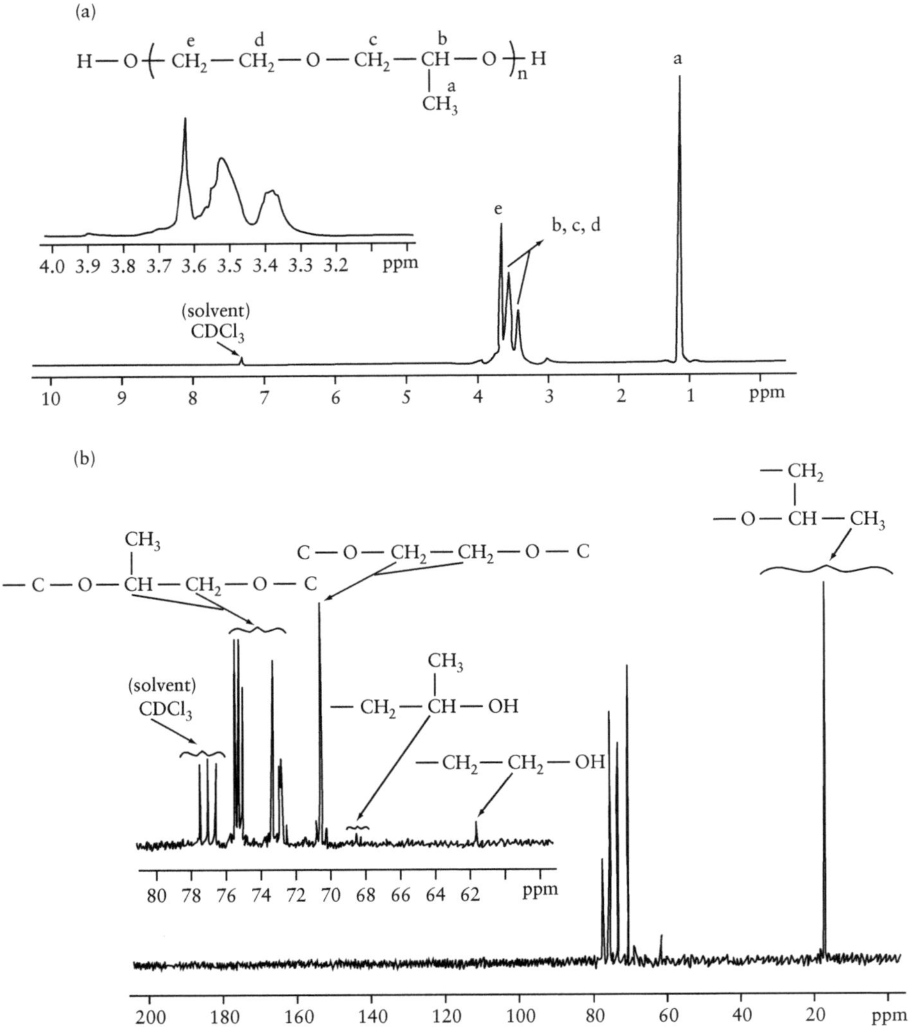

Figure 3.15: ^{1}H-NMR a) and ^{13}C-NMR b) spectra of virgin polyol. Reproduced with permission from M.M. Alavi Nikje, M. Nikrah and M. Haghshenas, *Polymer Bulletin*, 2007, **59**, 1, 91.. ©2007, Springer [5].

In Figure 3.16, the similarity in chemical structure of the recovered polyol to that of the virgin polyol is clear. However, there were various additional peaks in the ^{1}H-NMR and ^{13}C-NMR spectra of recovered polyols in comparison with the virgin polyol. In ^{13}C-NMR spectra, signals at 110–150 ppm were related to carbon atoms derived from the starting isocyanate. Additional peaks at 40, 114.5, 130.5, 129 and 146 ppm in the ^{13}C-NMR spectrum of the recovered polyol (Figure 3.17) were caused by aromatic diamine compounds. Also, the weak signals at 63.6 and 64.6 ppm were the typical bands of carbons bearing hydroxyl end-groups related to the transesterification byproducts obtained by substitution of the polyol by glycerin molecules.

In addition, no absorption band was observed at 156 or 164 ppm (to demonstrate urethane and urea carbonyl bonds, respectively). Thus, complete recovery of the polyol under MW irradiation was confirmed.

Figure 3.18 indicates that the peaks observed at 3.64, 6.5–6.6 and 6.8–6.9 ppm in the 1H-NMR spectrum of the recovered polyol corresponded to the aromatic diamine compounds remaining in the upper-phase, as well as peak areas for hydrogen atoms. Integration of contaminants and recovered polyol corresponding to hydrogen peak areas indicated polyol contamination to be 4%.

Figure 3.19 demonstrates the ^{1}H-NMR and ^{13}C-NMR spectra of the recovered polyol at 220 °C. All spectral regions are quite similar to those shown in Figure 3.16.

Application of MW irradiation accelerates glycolysis reactions in comparison with conventional methods. In order to confirm this hypothesis, we investigated a typical reaction at 180 °C with a conventional heating method, determined the time required for perfect foam dissolution, and compared it with the MW method. The geometry and particle size of the foam, the solvent, catalyst (and its concentration) and the solvent:foam ratio in both methods were quite similar. Obtained results indicated the ability of MW to decrease the reaction time from 29 min to 92 s in a KOH-accelerated reaction and from 42 min to 81 s in a NaOH-catalysed process (Figures 3.20 and 3.21, respectively).

3.5 Recycling of flexible foam wastes received from an auto seat production plant under microwave irradiation and split-phase conditions

3.5.1 Introduction and background

In the glycolysis reaction, the PU and polyurea functional groups in the polymer chain are degraded by successive transesterification reactions towards these functional groups with low-weight glycols in the presence of a moderately basic catalyst. Although the polyol was recovered and had a similar chemical structure to that of the virgin polyol, they were prepared in a liquid mixture of products containing hydroxyl-reactive groups. Because of the lower quality of the recycled polyol in

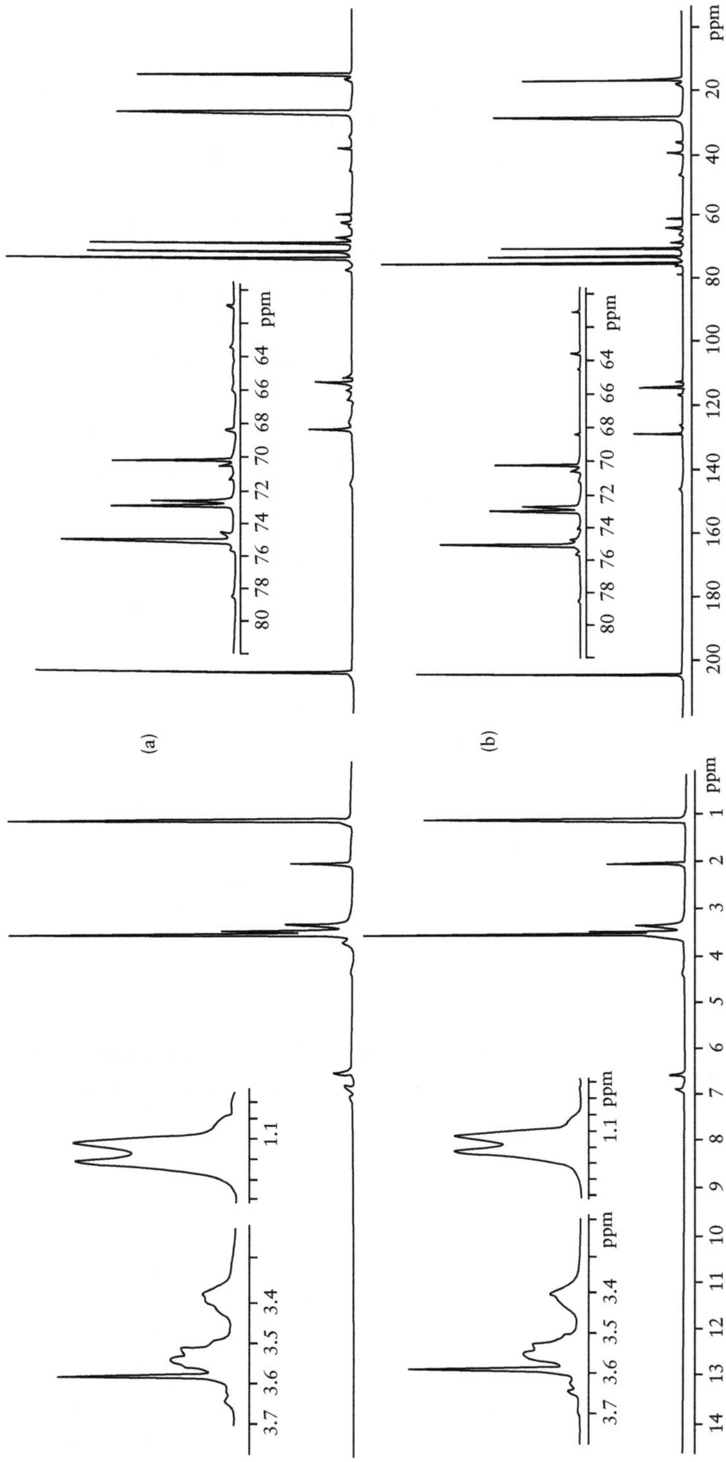

Figure 3.16: ^{1}H-NMR and ^{13}C-NMR spectra of recovered polyol at complete reaction time (30 s) in the presence of, a) NaOH and b) KOH at 160 °C. Reproduced with permission from M.M. Alavi Nikje, M. Nikrah and M. Haghshenas, *Polymer Bulletin*, 2007, **59**, 1, 91. ©2007, Springer [5].

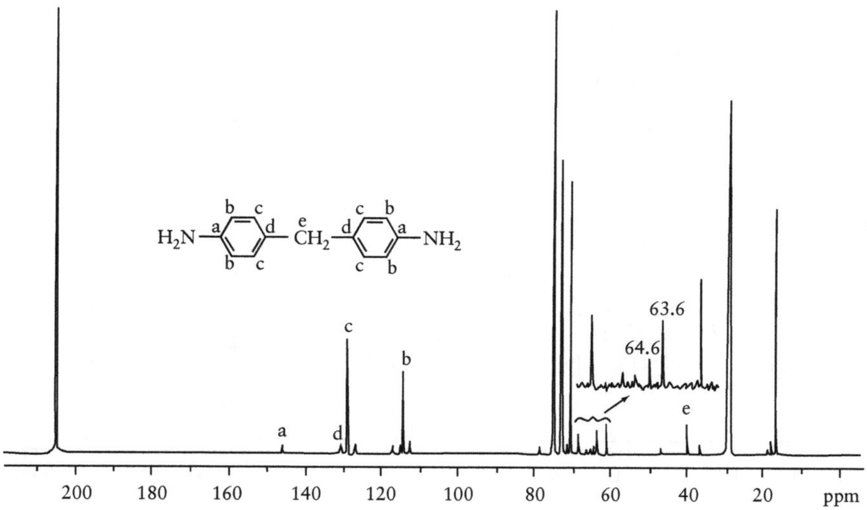

δ in ppm	Carbon
40	e
114.5	b
130.5	d
129	c
146	a

Figure 3.17: Regions associated with the transesterification byproducts remaining in the upper-phase as indicated in the ^{13}C-NMR spectrum of the recovered polyol. Reproduced with permission from M.M. Alavi Nikje, M. Nikrah and M. Haghshenas, *Polymer Bulletin*, 2007, **59**, 1, 91. ©2007, Springer [5].

comparison with that of the virgin polyol, they can be used only to produce medium-quality products or be blended with raw materials. Upon use of an excess amount of glycolysis agent (much larger than the stoichiometric quantity), the reaction product separated into two phases, whereas the upper-phase was formed mainly by the recovered polyol from the PU. In this study, the split-phase glycolysis reactions of MDI and polyether polyols-based flexible PUF were conducted with DEG and different basic catalysts in the presence of MW irradiations to achieve high-quality recycled polyols. Reactions were investigated and the obtained products were characterised using analytical and spectroscopic methods [6].

3.5.2 General procedure of the recycling process

Formulated flexible cold-cure PUF received from an industrial production plant making RIM parts was recycled in the split-phase condition. In the foam

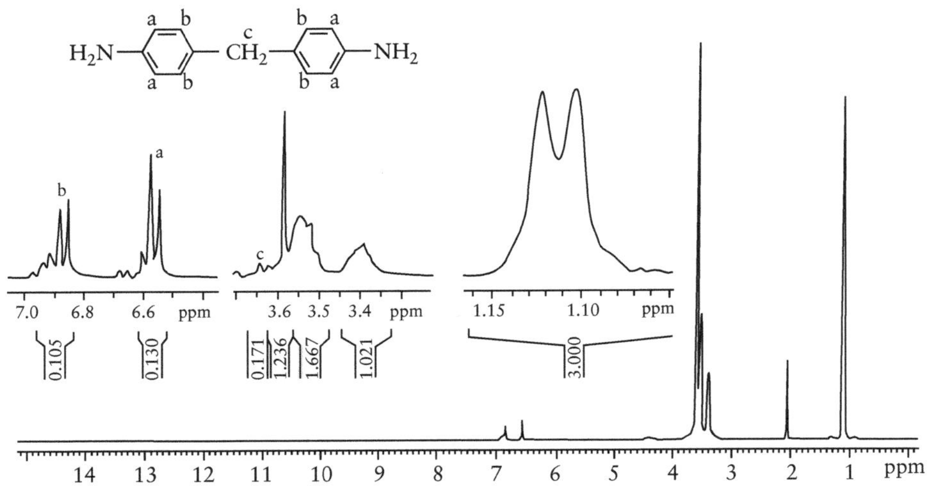

Figure 3.18: ^{1}H-NMR spectral regions associated with transesterification of byproducts (aromatic diamine compounds) remaining as contaminants in the upper-phase. Reproduced with permission from M.M. Alavi Nikje, M. Nikrah and M. Haghshenas, Polymer Bulletin, 2007, 59, 1, 91. ©2007, Springer [5].

formulation, the virgin polyol was Daltoflex® EC 20,240 and the isocyanate component was SUPRASEC® 2027 comprising MDI. The prefix 'virgin' was used to differentiate the polyol used first from the recycled polyol. Chemical structures of glycolysed products were studied by FTIR, ^{1}H-NMR and ^{13}C-NMR.

A well-specified amount of the base catalyst was dissolved in DEG and employed as the PU-destroying agent. Cold-cure foam waste and the formed destroying solvent were poured in a 100-ml Pyrex™ beaker. To ensure that all reactions were similar in all experiments, optimal immersion of PUF in the glycolysing agent was attempted. The beaker was placed in a domestic MW oven for selected times until perfect dissolution of the foam was reached. All reactions were done under atmospheric pressure. Disappearance of foam was selected as the glycolysis time. Neither of these reactions were completed at the glycolysis time, so the process was extended 10, 20, and 30 s after complete dissolution of the foam. Conversely, for evaluation of the different MW-power effects on glycolysis reactions, reactions were carried out at various power values. We used power values of 180, 300, 450, 600 and 900 W. At the end of the reaction, the mixture was removed from the MW oven, filtered for isolation of solid contaminants and un-reacted foams, and allowed to separate into two split-phases. After centrifugation of the upper-phase (recycled polyol), its specifications were characterised and compared with those of the virgin polyol.

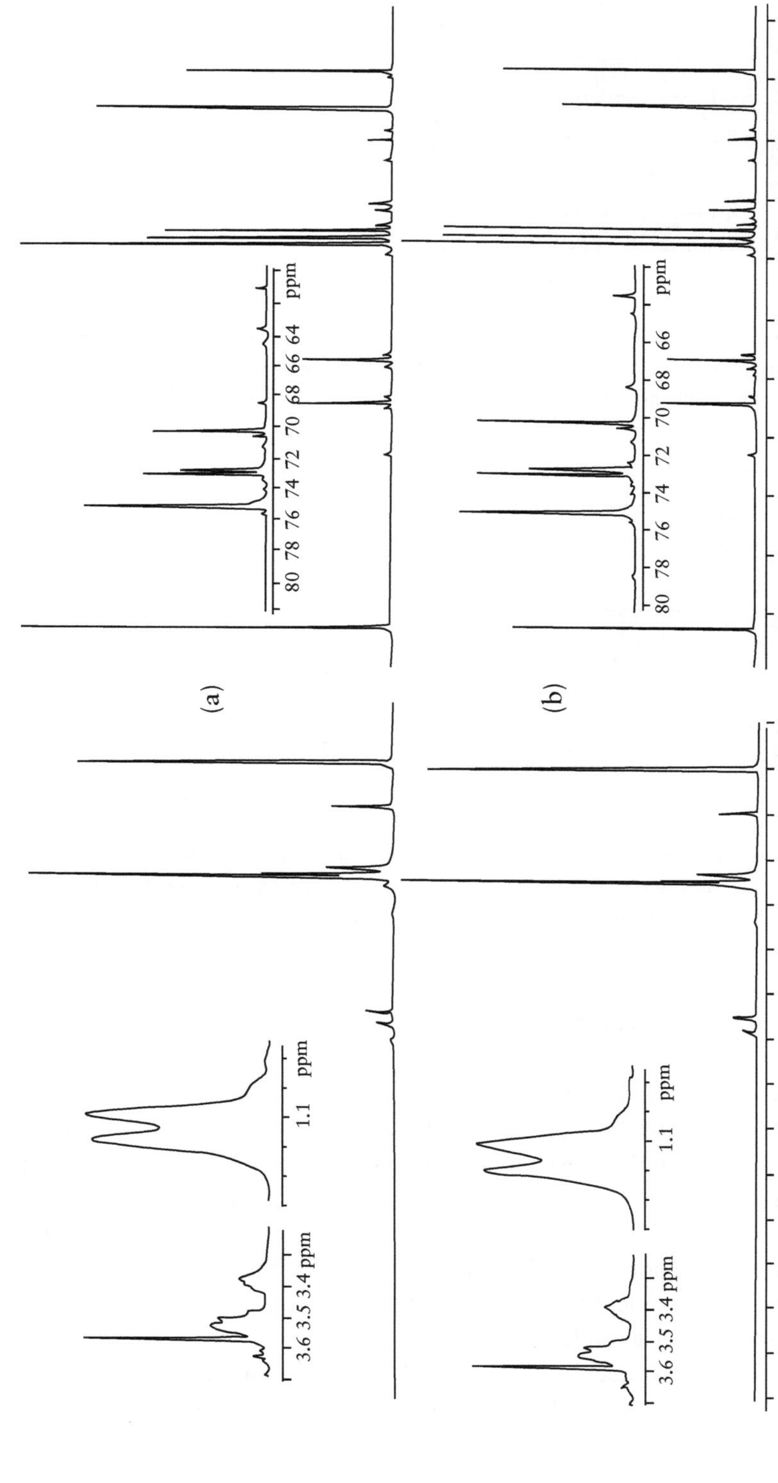

Figure 3.19: ^{1}H-NMR and ^{13}C-NMR spectra of the recovered polyol at 220 °C, (a) in the presence of NaOH immediately after perfect foam dissolution and (b) at 10 s reaction time in the presence of KOH. Reproduced with permission from M.M. Alavi Nikje, M. Nikrah and M. Haghshenas, *Polymer Bulletin*, 2007, **59**, 1, 91. ©2007, Springer [5].

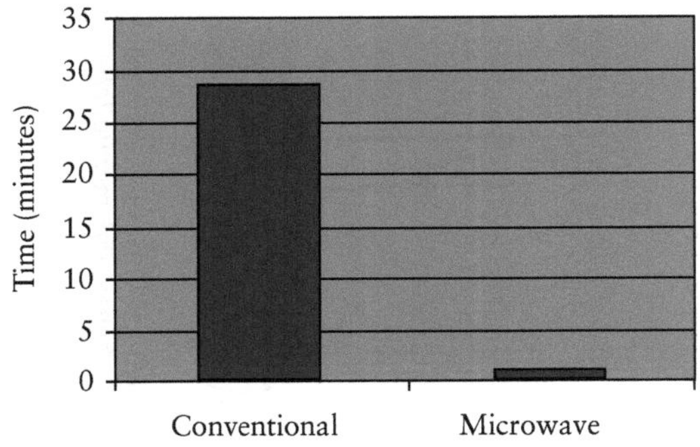

Figure 3.20: Comparison of times for complete foam dissolution in the presence of KOH at 180 °C under MW and conventional heating conditions. Reproduced with permission from M.M. Alavi Nikje, M. Nikrah and M. Haghshenas, *Polymer Bulletin*, 2007, **59**, 1, 91. ©2007, Springer [5].

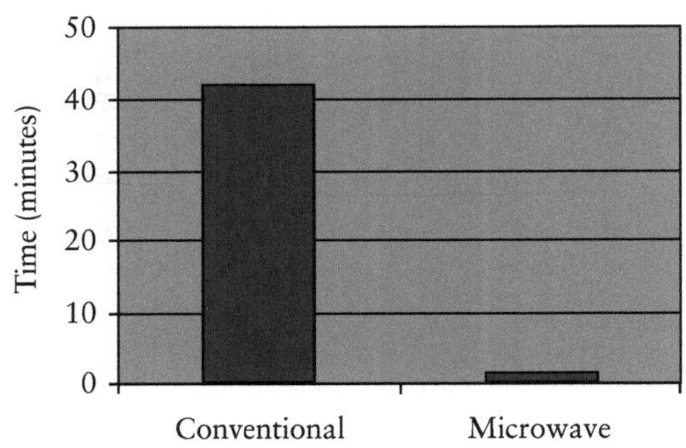

Figure 3.21: Comparison of times for complete foam dissolution in the presence of NaOH at 180 °C under MW and conventional heating conditions. Reproduced with permission from M.M. Alavi Nikje, M. Nikrah and M. Haghshenas, *Polymer Bulletin*, 2007, **59**, 1, 91. ©2007, Springer [5].

3.5.3 Interpretations of obtained data

To investigate the effect of MW irradiations on the glycolysis of PUF foams, we examined the dependence of different reaction times, MW power, and various basic catalysts on glycolysis reactions. We found that dissolution of PUF in the destroying solvent without a catalyst was extremely difficult. Hence, various catalysts were employed in combination with a solvent system to decrease the required dissolution time. Solvent systems (containing DEG and different catalysts) were evaluated to find the best dissolution time. For comparison, the glycolysis of foams was investigated in the absence of base catalysts.

Data were collected as described in Figure 3.22. As shown in the figure, the best results were obtained by using NaOH and KOH as catalysts. Conversely, when the

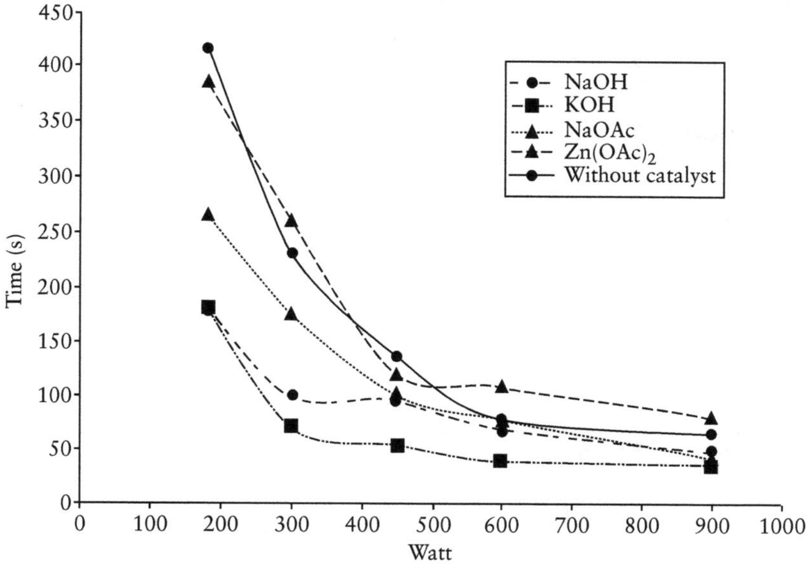

Figure 3.22: Comparison of foam dissolution times for different MW powers in the presence of various catalysts. Reproduced with permission from M.M. Alavi Nikje and M. Nikrah, *Polymer – Plastics Technology and Engineering*, 2007, **46**, 4, 409. ©2007, Taylor & Francis [6].

reaction was investigated in the absence of a catalyst, dissolution times were long in comparison with catalytic reactions. There were considerable differences between the dissolution times in catalysed reactions with NaOH and KOH in comparison with those using sodium acetate (NaAc) and zinc acetate [$Zn(OAc)_2$]. Also, the physical state of obtained products using various catalysts was different. For example, recycled polyols by $Zn(OAc)_2$, NaAc, as well as the polyols obtained in the absence of base catalysts, due to imperfect degradation of urethane chains, had higher viscosities.

Figure 3.23 shows the FTIR spectra of the starting polyol used in PUF production. Intense stretching vibrations of the aliphatic ether group at 1,115 cm^{-1} that overlapped the band produced by CO groups related to hydroxyl end-groups and stretching vibrations of these hydroxyl groups were found at 3,482 cm^{-1}. Absorption bands in the region of 2,970–2,868 cm^{-1} (due to stretching vibrations of CH bonds in aliphatic carbons) and in 1,456 and 1,374 cm^{-1} (characteristic of bending vibrations of methylene groups in the polyol chain) were observed.

Comparison of the FTIR spectra of recycled polyols at different MW power values with the virgin polyol indicated that the chemical structures of recovered polyols by glycolysis in the presence of KOH and NaOH as catalysts were quite similar to those of the virgin polyol (Figures 3.24 and 3.25, respectively). This finding was indicated by absorption bands characteristic of the starting polyol in the upper- phase.

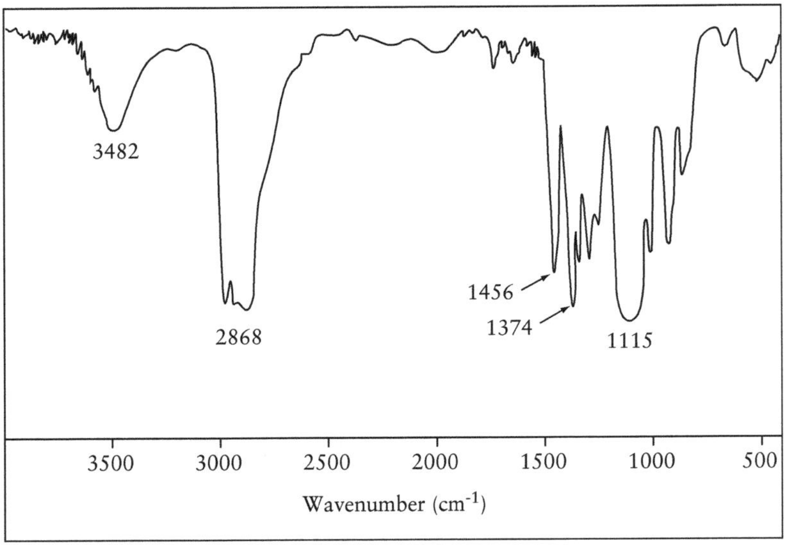

Figure 3.23: FTIR spectra of virgin polyol. Reproduced with permission from M.M. Alavi Nikje and M. Nikrah, *Polymer – Plastics Technology and Engineering*, 2007, **46**, 4, 409. ©2007, Taylor & Francis [6].

Obtained upper-phases were polluted in small quantities by the main products of the lower-phase, which made the spectra slightly different from those of the pure polyol. The glycol dissolved in the upper-phase did not affect absorption bands in an appreciable way because its molecular structure was similar to that of the constituent units of one of the blocks of the polyol, except for the increase of the quantity of hydroxyl groups (absorption band at 3,368 cm^{-1}). The new absorption bands at 1,629 and 1,515 cm^{-1} in Figure 3.24 and 1,623 and 1,523 cm^{-1} in Figure 3.25 corresponded to transesterification and decarboxylated products, which could have been related to the primary amines formed (bending vibrations of NH) due to partial solubility in the upper-phase, which polluted the recycled polyol.

Bands related to stretching vibrations of the CO bond in urethane groups were found at around 1,690–1,749 cm^{-1}. Hence, the bands observed at 1,730 and 1,729 cm^{-1} in Figure 3.24 could be assigned to urethane groups, and their intensity decreased with increasing of MW power.

Figures 3.25 and 3.26 show the FTIR spectra of the recycled polyol at different MW power values in the presence of NaAc and $Zn(OAc)_2$ as base catalysts, respectively. All spectral regions described above were repeated in these spectra, except for the noticeable decrease in intensity of the urethane bond at 1,730 cm^{-1} with increasing of MW power. This finding demonstrated the inability of NaAc and $Zn(OAc)_2$ to breakdown completely polymeric chains in a short time even at strong MW power (900 W). Difficulty in carrying

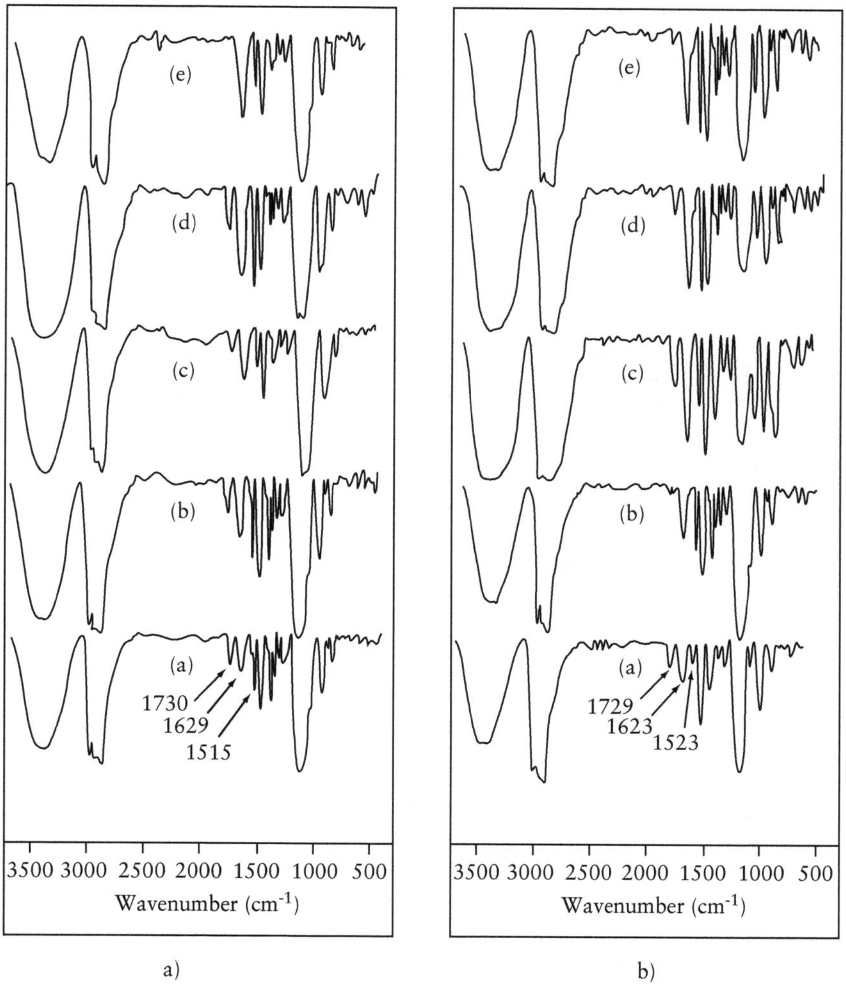

Figures 3.24: Comparison of FTIR spectra of the recycled polyol in the presence of a) KOH and b) NaOH respectively, as base catalysts 30 s after perfect dissolution at MW power values of (a) 180; (b) 300; (c) 450; (d) 600; and (e) 900 W. Reproduced with permission from M.M. Alavi Nikje and M. Nikrah, *Polymer – Plastics Technology and Engineering*, 2007, **46**, 4, 409. ©2007, Taylor & Francis [6].

out reactions without catalysts was due to the requirement of high activation energy, so the reactions were barely carried out at 180 W and the obtained polyols were not of appropriate quality. Thus, uncatalysed reactions could not be completed in a short reaction time (30 s).

A urethane bond at 1,729 cm^{-1} with high intensity in the FTIR spectra of obtained polyols from the upper-phase confirmed non-completed dissolution of polymeric chains (Figure 3.27).

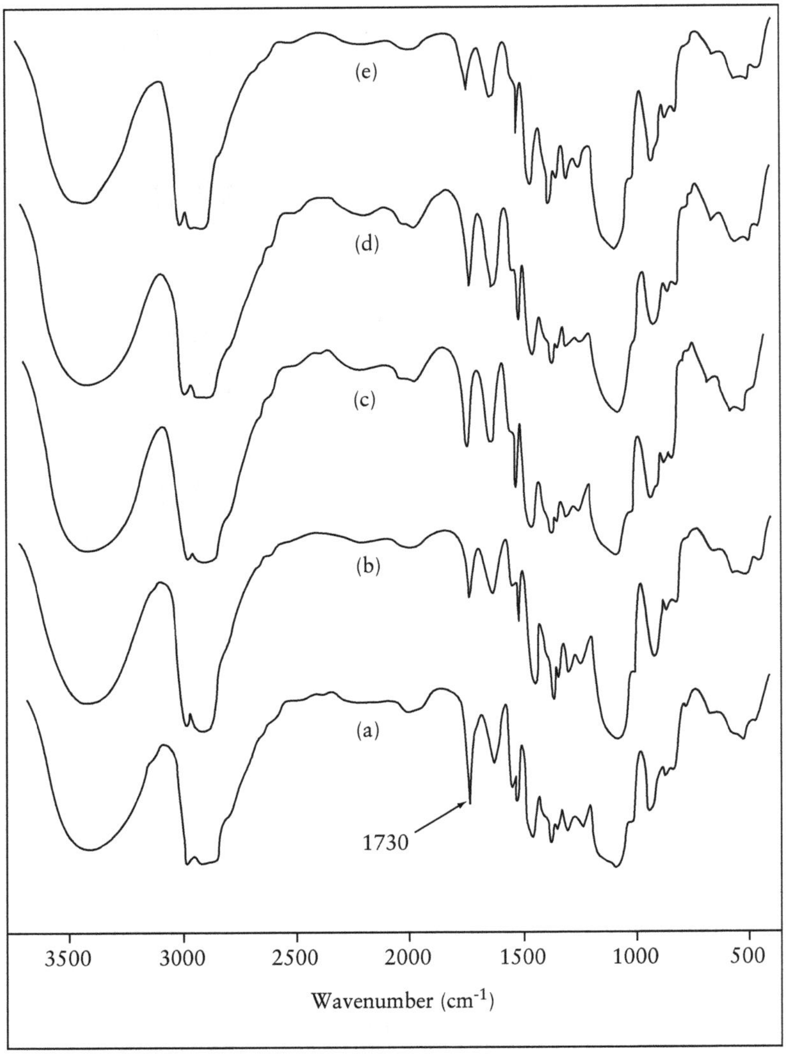

Figure 3.25: FTIR spectra of recycled polyols in the presence of NaAc as a catalyst at MW powers of (a) 180; (b) 300; (c) 450; (d) 600; and (e) 900 W. Reproduced with permission from M.M. Alavi Nikje and M. Nikrah, *Polymer–Plastics Technology and Engineering*, 2007, **46**, 4, 409. ©2007, Taylor & Francis [6].

For further comparison and evaluation, the NMR spectra of the recycled polyol with NaOH as a sufficient catalyst for the glycolysis process under MW irradiation and NaAc as a catalyst of lower suitability were evaluated. Figure 3.28 shows the ^{1}H-NMR and ^{13}C-NMR spectra of the virgin polyol. In the ^{1}H-NMR spectrum, the peak in the region 1.1 ppm is related to methyl-equivalent hydrogen atoms, and

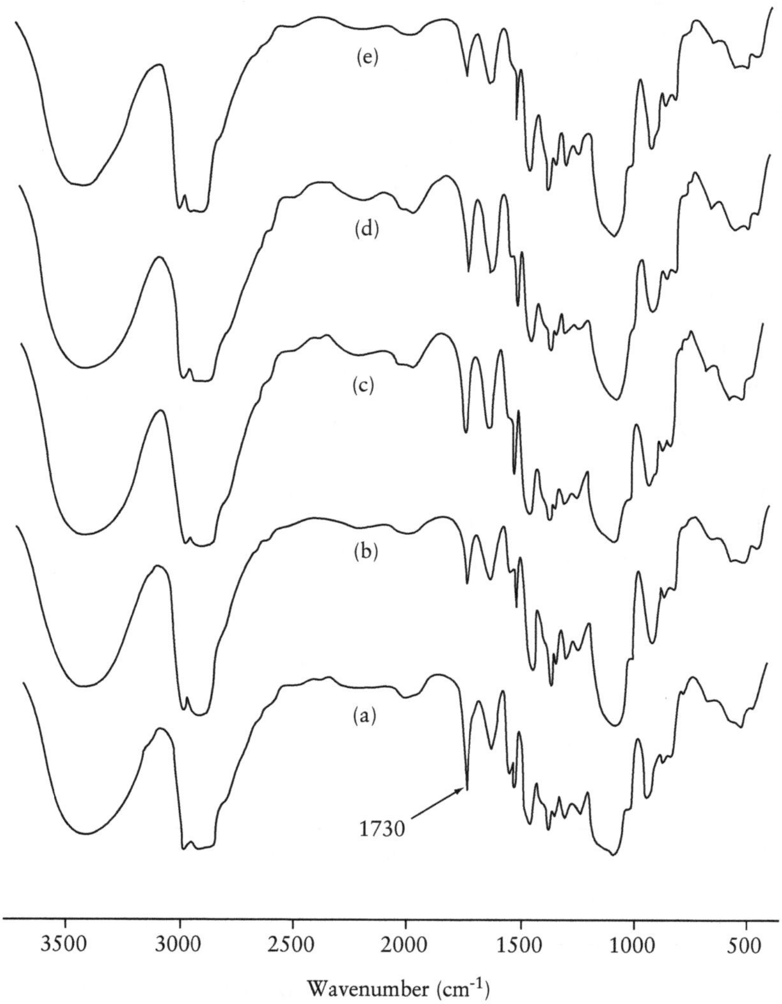

Figure 3.26: FTIR spectra of recycled polyols in the presence of $Zn(OAc)_2$ as a base catalyst at a MW power of (a) 180; (b) 300; (c) 450; (d) 600; and (e) 900 W. Reproduced with permission from M.M. Alavi Nikje and M. Nikrah, *Polymer–Plastics Technology and Engineering*, 2007, **46**, 4, 409. ©2007, Taylor & Francis [6].

peaks in the region 3.0–4.0 ppm corresponding to hydrogen atoms and their carbon atoms connected directly to the oxygen atom.

Figure 3.29 shows a comparison of the ^{1}H-NMR spectra of a virgin polyol with a) the recycled product in the upper-phase using NaOH, b) NaAc and c) as a catalyst. Chemical structure of the obtained polyol using NaOH was too similar to that of the virgin polyol, except that the peaks had very low intensity in the 7.0–7.3 ppm region, which were related to the hydrogen atoms of the aromatic byproducts derived from

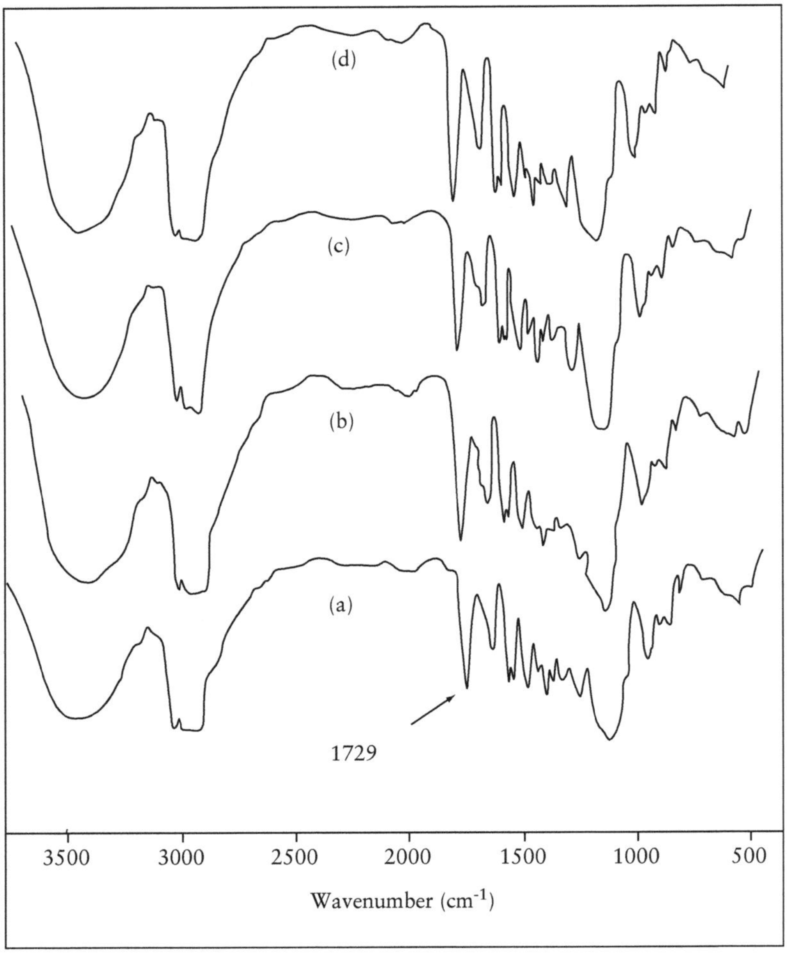

Figure 3.27: FTIR spectra of recycled polyols in the absence of a catalyst at MW power of (a) 300; (b) 450; (c) 600; and (d) 900 W. Reproduced with permission from M.M. Alavi Nikje and M. Nikrah, *Polymer–Plastics Technology and Engineering*, 2007, **46**, 4, 409. ©2007, Taylor & Francis [6].

the starting isocyanates that partially remained in the recycled polyol. Conversely, the ^{1}H-NMR spectrum of the polyol obtained with NaAc was more different in comparison with the virgin polyol. The spectral region at 3.0–4.0 ppm, due to non-glycolysed polymeric chains, was more crowded.

The signal in the 31 ppm region in the ^{13}C-NMR spectrum of the recovered polyol in the presence of NaAc was related to the carbon atom of the methyl group in oligomeric chains (Figure 3.30c). Also, the peak in the 207 ppm region corresponded to the carbon of urethane functional groups, which could not be

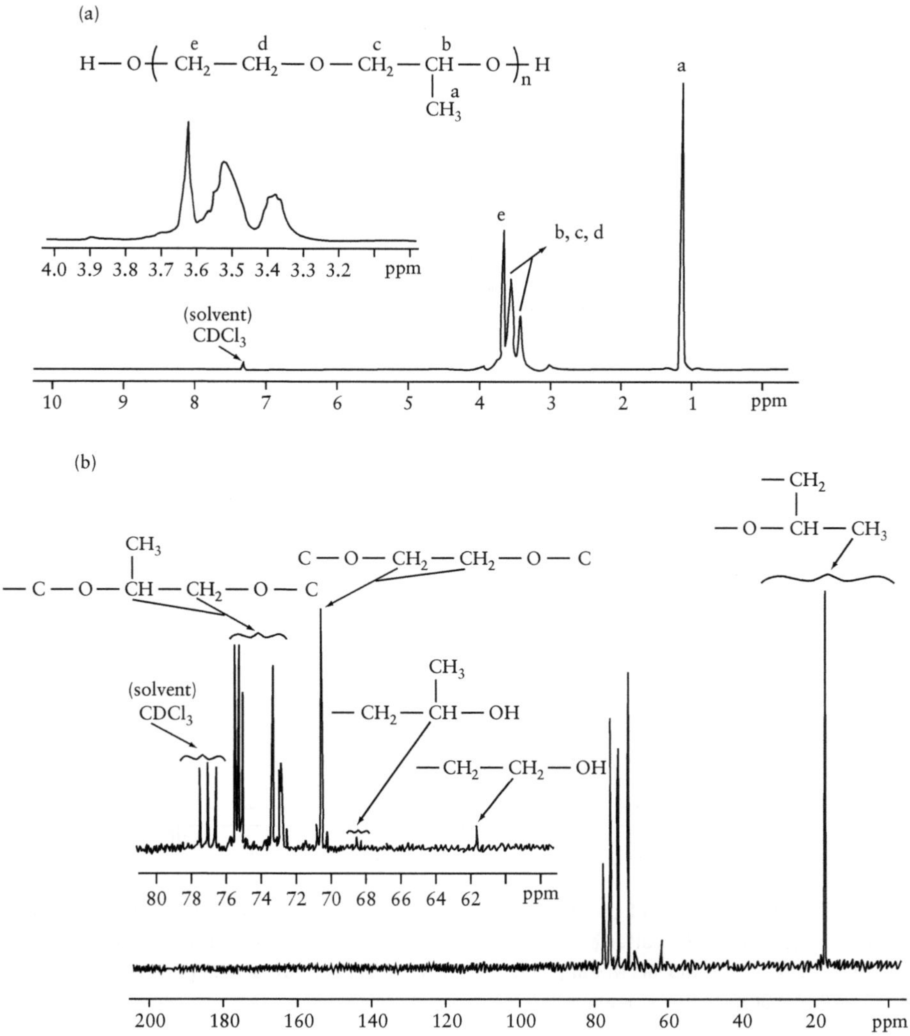

Figure 3.28: ^{1}H-NMR and ^{13}C-NMR spectra of the virgin polyol. Reproduced with permission from M.M. Alavi Nikje and M. Nikrah, *Polymer–Plastics Technology and Engineering*, 2007, **46**, 4, 409. ©2007, Taylor & Francis [6].

degraded at 30 s. In the ^{13}C-NMR spectrum of recycled polyol with NaOH (Figure 3.30b), due to the ability of this catalyst to completely degrade urethane functional groups, signals at 31 and 207 ppm were not observed. According to these results, MW could be employed as an energy- strong source for glycolysis processes that gave rise to reactions carried out in a very short time. Also, an appropriate base catalyst in the solvent system was needed to achieve a high-quality polyol.

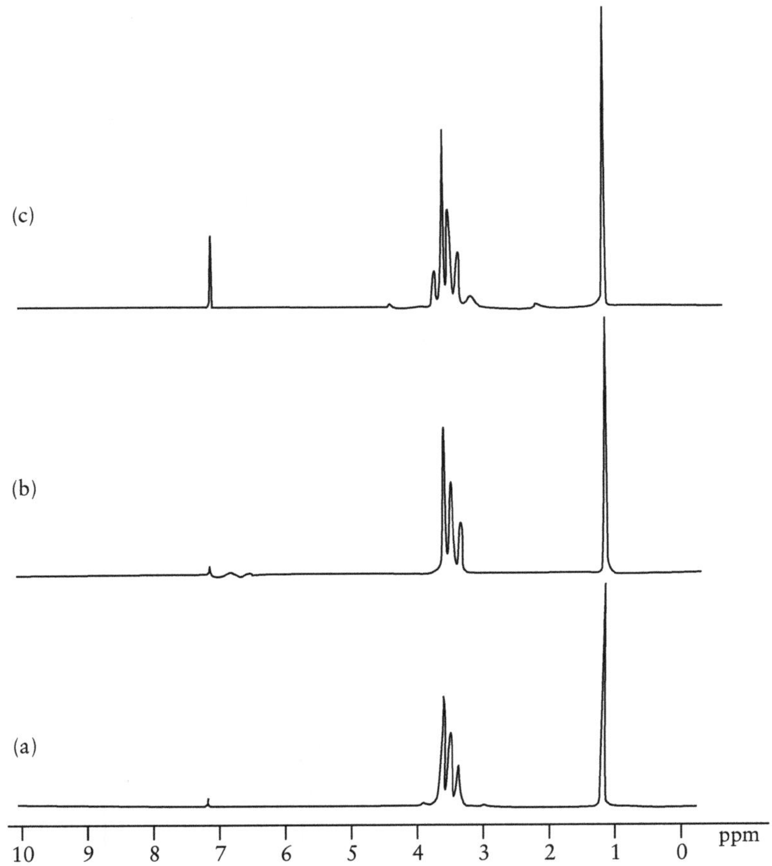

Figure 3.29: ^{1}H-NMR spectra of the virgin polyol (a), the recycled polyol with NaOH, (b) NaAc and (c) at 900 W and 30 s after perfect dissolution of foam. Reproduced with permission from M.M. Alavi Nikje and M. Nikrah, *Polymer–Plastics Technology and Engineering*, 2007, **46**, 4, 409. ©2007, Taylor & Francis [6].

3.6 Introduction of hydroglycolysis as a benign and environmentally friendly method for recycling of flexible polyurethane foam wastes under microwave irradiation

3.6.1 Introduction and background

As mentioned in Chapter 2, the hydrolysis process is carried out under high temperature and pressure, and leads to a simple product mixture. Glycolysis reactions are done under mild conditions, but several reaction products are produced. Therefore, a combined hydroglycolysis process is beneficial from economic and environmental viewpoints because it uses less petrochemical solvent and produces fewer byproducts.

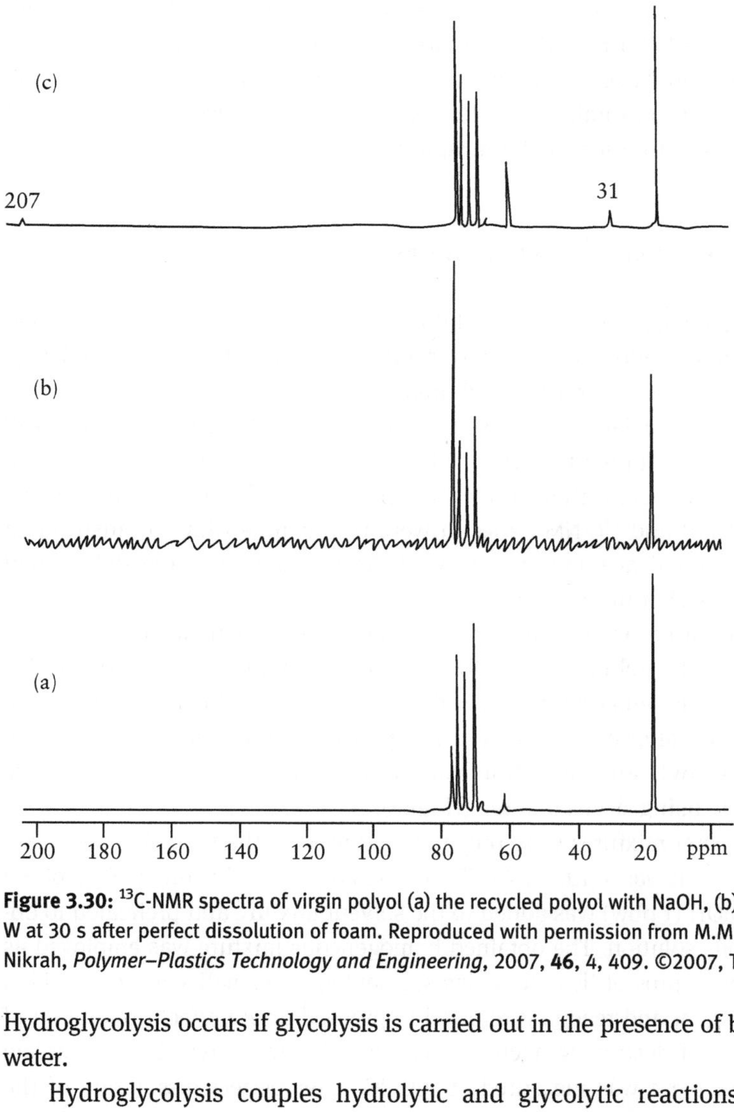

Figure 3.30: ^{13}C-NMR spectra of virgin polyol (a) the recycled polyol with NaOH, (b) NaAc and (c) at 900 W at 30 s after perfect dissolution of foam. Reproduced with permission from M.M. Alavi Nikje and M. Nikrah, *Polymer–Plastics Technology and Engineering*, 2007, **46**, 4, 409. ©2007, Taylor & Francis [6].

Hydroglycolysis occurs if glycolysis is carried out in the presence of base catalysts and water.

Hydroglycolysis couples hydrolytic and glycolytic reactions to degrade PU chains. That is, water in the solvent mixture as a co- reagent causes hydrolysis reactions to take place beside glycolysis reactions. CO_2 obtained *in situ* by the reaction of water with MDI is the chief blowing agent for all commercially manufactured flexible PUF. The reaction of water with MDI provides most of the gas formed to be used to expand the urethane polymer. Participation of water in the polymerisation reaction leads to an expanded cellular urethane polymer.

Another product of this reaction is diamine, which reacts immediately with additional isocyanate and leads to formation of urea bonds in the PU structure. The reaction of urea and urethane with water in the hydrolysis method leads to the formation of diamine and CO_2.

In this work, chemical recycling of flexible PUF wastes was achieved through hydroglycolysis under MW irradiation at atmospheric pressure and 160 °C. Mixtures of glycerin, water, and NaOH were employed as hydroglycolysis agents. This research was conducted to examine the effect of water on the glycolysis reactions of flexible PUF under MW irradiation at atmospheric pressure [7].

3.6.2 General procedure of the recycling process

For the recycling process, industrial samples of flexible cold-cure PUF were derived from the reaction of Daltoflex® EC 20,240 (formulated virgin polyol) and MDI. Isocyanate was SUPRASEC® 2027, a MDI- based prepolymer.

Experiments were carried out in a Milestone MicroSYNTH 'NP' Ethos 1600 MW oven under atmospheric pressure. FTIR, ^{1}H-NMR and ^{13}C-NMR spectroscopic methods were used for characterisation of recycled products and comparison with the virgin polyol. ^{1}H-NMR and ^{13}C-NMR spectra were run using a CRX 300 instrument (Bruker) and deuterated acetone as a solvent. Water contents were determined using a DL 18 Karl Fischer titrator (Mettler).

M_WD were measured using a 6A GPC instrument (Shimidzu) with a particular column type [Bruker PL gel (5 µl)] at 1,000 A° and a column length of 30 cm. The mobile-phase was THF, temperature was 408 °C, flow rate was 1 cm^3/min, injection volume was 100 µl, calibration was by using polystyrene standards (950, 2,200, 5,100 and 1,1600 Mw), and detection was at RI conditions. Viscosities were measured at 25 °C using a viscometer (Haake).

Glycerin and water mixtures were prepared and used as the hydroglycolysis solvent. The water content was 5, 10, 15, 20, 25, 30, 35 and 40% of the total weight of the solvent system. NaOH (1 pbw) was added to the solvent mixture and preheated to obtain a homogeneous solution. The obtained homogeneous mixture was employed as a destroying agent. Scraps of flexible foams segmented into small sizes were added to the destroying agent and reactions allowed to proceed until perfect dissolution of the foam. The ratio of destroying agent to foam was 1:1. In all reactions, complete dissolution time of foam was measured at 160 °C. After perfect dissolution of the foam at a certain time, the container holding the reaction mixture was removed from the MW oven and allowed to cool at RT. Contents was filtered and delivered until they split into two distinct phases. The obtained product was characterised by spectroscopic methods, and data compared with data obtained from an authentic polyol.

3.6.3 Interpretations of obtained data

To prevent undesired reactions, hydroglycolysis reactions were undertaken at 160 °C. It has been shown that, at high temperatures (>200 °C), unwanted secondary reactions become competitive with respect to transesterification reactions, and pyrolysis products cannot be excluded. Conversely, at low temperatures, catalytic activity is reduced, resulting in slowing of the dissolution rate and a longer reaction time. Obtained results revealed that the hydrolysis and glycolysis of flexible PUF could be carried out at 160°C using MW irradiation at atmospheric pressure. Also, use of the MW method with the hydroglycolysis method provided the opportunity to use ≤40% water in the solvent mixture while maintaining a reasonable reaction time. The hydroglycolysis reactions of flexible PUF were carried out at a split-phase condition, just like in glycolysis reactions. The upper-phase consisted of recovered polyol and the lower-phase included aromatic diamines, transesterification byproducts, excess unreacted glycerin, and water.

Addition of water as the second reagent in the hydroglycolysis agent slowed dissolution of the foam (only 20% of the foam was dissolved after 15 min), but it is a potential low-cost replacement for ≤40% of glycerin because reasonable dissolution times were still achievable when compared with the 10 h needed for conventional heating.

Conversely, the water content in recovered polyols indicated the overall trend of increasing water amounts in the solvent mixture. These recovered polyols could be used for the manufacture of new water-blown PUF, so small amounts of water should not be problematic. Flexible foam polyols are usually supplied with water content <0.1% to ensure consistency and flexibility (even though water is added later when producing foam).

Nevertheless, the water concentration resulting from hydroglycolysis is of no concern because water can be removed readily in commercial practice. ^{1}H-NMR and ^{13}C-NMR spectra of recovered polyol were compared with those of the virgin polyol. Four characteristic peaks at 1.1 (3H, – CH_3) and 3.3–3.6 (6H, – CH_2–O – and 1H, – CH–O–) ppm in the ^{1}H-NMR spectra of recovered and virgin polyols demonstrated their structural similarity. However, due to the partial solubility of aromatic diamine [4,4′-methylenedianiline (MDA)] in the recovered polyol as a contaminant, additional peaks at 3.64, 6.5–6.6 and 6.8–6.9 ppm were observed in the ^{1}H-NMR spectrum of the recovered polyol that were similar to the sample data from Section 3.5. Also, a weak peak at 4.4 ppm corresponded to water. In the ^{13}C-NMR, the characteristic peaks were: 17.3 [–OCH(CH_2)–CH_3], 61.1 (– CH_2–CH_2–OH), 68.5 [–CH_2 – CH(CH_3)–OH], 70.2–70.7 (C–O – CH_2–CH_2–O–C) and 72.5–75.3 [C–O–CH (CH_3)–CH_2–O–C] ppm. Peaks in the regions 40, 114.5, 129, 130.5, and 146 ppm in the ^{13}C-NMR spectrum of the recovered polyol were related to the carbons originating from MDA as a contaminant.

Comparison of the FTIR spectra of recovered polyols obtained by destroying agents containing 5 and 20% water at a given reaction time after complete dissolution of the foam indicated that the ability of the destroying agent for breaking down PU linkages decreased with increasing water content in the solvent mixture from 5 to 20%.

Incomplete scission of PU chains at an absorption band at 1,728 cm^{-1} corresponded to the stretching vibrations of the carbonyl groups in urethane linkages in the FTIR spectrum of the recovered polyol by a destroying agent containing 20% water. Distinctive absorption bands at 1,115, 3,482, 2,868, 1,456, and 1,374 cm^{-1} in the FTIR spectrum of Daltoflex® EC 20,240 as the virgin polyol were related to the stretching vibrations of ether (–C–O–C–) groups, – OH groups related to hydroxyl end-groups, as well as the – CH bonds and bending vibrations of – CH_2 and – CH_3 groups in the polyol chain, respectively. These absorption bands were present in the FTIR spectra of recovered polyols except for two new absorption bands at 1,627 and 1,515 cm^{-1}, which corresponded to the bending vibrations of the amine groups of aromatic diamines formed during the process (trace amounts of which were dissolved in the recovered polyol). An absorption band at 1,728 cm^{-1} in the recovered polyols demonstrated the incomplete breakdown of urethane bonds and, therefore, the presence of urethane-containing oligomers.

The M_WD and viscosity of the recycled product are critical to its application, so they were measured in recycled polyols and data were compared with those of the virgin polyol.

To reduce the amine content of recycled products, a glycerin:water mixture was employed as the glycolysing agent and NaOH as a catalyst, and experiments were followed in the absence of an amine decomposer.

According to the results obtained from GPC chromatograms, the starting polyol was recovered completely as the major product in the upper-phase, and with similar elution times and characteristics to those of the virgin polyol. Conversely, absence of additional peaks in the GPC of recovered polyols in the upper-phase indicated the complete recovery of polyols without product contamination with the bottom liquid-phase having a low-Mw.

In another attempt, the viscosity of recycled polyols was obtained and results compared with the viscosity of the virgin polyol: and results were slightly similar. The viscosity of Daltoflex® EC 20,240 was 1,280 cps, whereas the recycled polyols had increased viscosity in comparison with the virgin polyol (probably due to additives employed for formulation of a polyol on an industrial scale). That is, the viscosity of the recycled polyol decreased with increasing water content and increasing reaction time due to successful scission of PU chains. Results obtained from viscosity studies confirmed GPC, NMR and IR data, and suggested complete degradation of urethane bonds.

3.7 All green solvents mixture as a novel glycolysis agent for chemical recycling of flexible polyurethane foam under microwave irradiation

3.7.1 Introduction and background

To continue the experiments reported above, we investigated the performance of a glycerin/sorbitol/water system combined with NaOH as a new 'green' destroying agent in the hydroglycolysis reactions of flexible PUF. In this research, an environmentally friendly and convenient method was used for recycling of flexible PUF. A glycerin/sorbitol/water mixture was employed as a destroying solvent and NaOH as a catalyst. In all reactions, phase separation appeared after perfect digestion of foam. The upper-phase contained recycled polyol and the lower-phase was a brown liquid applied in rigid PUF formulations. Reactions were investigated using different ratios of glycerin/sorbitol/water. To save energy and enhance reaction rates, all reactions were carried out using a commercial MW oven [8].

3.7.2 Experimental section

3.7.2.1 Materials and instruments

Samples of flexible PUF were obtained from the flexible foam polyols Daltoflex® EC 20,240 and MDI-SUPRASEC® 5005.

Tensile properties were measured using an 1122 tensometer (5 KN; Instron), at a cross-head speed of 5 mm/min at RT, according to ASTM D638. Three specimens of each sample were tested. In these samples, the tensile trend was linear according to the force applied. Thus *E*-modulus results were obtained from the slope of stress–strain curves.

3.7.2.2 Hydroglycolysis

Scraps of flexible PUF (5 g) (segmented into small sizes), NaOH (1% *w/w*) and various systems in which glycolysis reactions were running, were placed in a two-necked flask, and then placed into a MW oven under atmospheric pressure at 180 °C and 800 W for different reaction times. When foam digestion was complete, the reaction mixture was removed from the MW oven and left to separate into two distinguishable phases. Finally, the upper-phase (recycled polyol) was characterised by spectroscopic methods.

3.7.2.3 Preparation of new rigid polyurethane foam using lower recycled phases

To use the lower-phase in rigid PUF formulations, this phase (which contained unreacted destroying solvents and some recovered amines [e.g., MDA)] directly was mixed in various ratios with virgin polyol (DALTOFOAM® TA 14,066; purchased from Huntsman, UT, USA), aliphatic amines, catalysts, surfactant and water. Virgin polyol in a blend with the lower-phase and additives mentioned above were mixed with MDI (SUPRASEC® 5005) in free-rise conditions for the manufacture of rigid PUF. For comparison, a sample of rigid PUF made of virgin polyol DALTOFOAM® TA 14,066 (100 parts) and MDI-SUPRASEC® 5005 (133 parts) was prepared.

3.7.3 Results and discussion

3.7.3.1 Characterisation of recovered polyols

As mentioned before, PUF dissolution in destroying solvents without catalysts was extremely difficult and basic catalysts (namely KOH or NaOH) were appropriate from the viewpoints of reaction rates and economics. Hence, we employed NaOH as the catalyst. To study the effect of sorbitol on recycling of flexible PUF, we investigated the dependence of reaction time *versus* sorbitol content. We found out that, with increasing sorbitol content, the reaction time increased. This effect was probably due to the increasing viscosity of the reaction media as well as difficulties in nucleophilic attack of the destroying agent to PU bonds. The increase in the sorbitol portion >50% of the destroying agent lowered the reaction times. From the FTIR spectra of recycled polyols and virgin polyols, the absorption band at 1,115 cm^{-1} corresponded to the aliphatic ether group of polyether polyol. Bending vibrations of methylene groups appeared in the polyol chain at 1,374 and 1,456 cm^{-1}, stretching vibrations of CH bonds in aliphatic carbons were found at 2,970–2,868 cm^{-1}, and stretching vibrations of hydroxyl groups were found at 3,482 cm^{-1}. Similar to our other experiments, ^{1}H-NMR spectra of the recovered polyol were compared with those for the virgin polyol. There were four characteristic peaks – at 1.1 ppm (3H, $-CH_3-$) and 3.3–3.6 ppm (6H, $-CH_2-O-$ and 1H, -CH-O-) – in the ^{1}H-NMR spectra of recovered and virgin polyols that demonstrated their structural similarity. The weak peak near 5.0 ppm was related to water. In the ^{13}C-NMR spectra, the characteristic peaks consisted of: 17.3 ppm [$-OCH (CH_2)-CH_3$], 61.1 ppm ($-CH_2-CH_2-OH$), 68.5 ppm [$-CH_2-CH (CH_3)-OH$], 70.2–70.7 ppm ($C-O-CH_2-CH_2-O-C$), and 72.5–75.3 ppm [$C-O-CH (CH_3)-CH_2-O-C$]. The presence of a contaminant in the recovered polyol has been proved by comparing carbon signals in ^{13}C-NMR spectra of recycled and virgin polyols, respectively. In this study, signals at 40, 114.5, 129, 130.5 and 146 ppm in ^{13}C-NMR spectrum of recovered polyol are related to the carbons originated from MDA as contaminant in the recovered polyol.

3.7.3.2 Preparation and characterisation of rigid polyurethane foams

An important observation in the reactions for preparation of rigid foams from virgin polyols and lower recyclate-phases was the compatibility of these two products that makes them useful in a similar formulation for manufacture of new rigid PUF.

Our data revealed that increasing the content of the lower-phase by ≤20% caused an increased density in the PUF prepared. An increase in lower-phase content >20% in a blend with a virgin polyol was not appropriate because of a reduction in the quality of the PUF formulation. In these conditions, determination of some of the parameters of the foaming process was not possible. In the meantime, factors in the foaming process (tack free, string gel, and free rise times) were diminished when lower recycled-phase content in the polyol blend increased. Conversely, the foaming process of blends containing high recyclate content was more exothermic, and heat release during the foaming process is greater with increasing recyclate content in the polyol blend.

According to the data obtained from the mechanical properties of prepared samples, tensile strength and elongation at break of formulated foams decreased with increasing content of the lower- phase (which contained sorbitol). Data comparison of the mechanical properties of foams made with virgin polyols or lower-phase from recycled foam confirmed that the foams prepared using the residue (lower-phase) of the glycolysis process were approximately similar to those prepared with virgin polyols. All samples prepared using the lower-phase of the glycolysis product were more rigid (lower elongation, bigger E-modulus values) than the one prepared from virgin polyol exclusively (functionality = 2). This phenomenon was due to increments in crosslinking densities dependent on sorbitol (functionality = 6). In addition, unreacted sorbitol made the foam more physically dense and caused E-modulus enhancement.

3.8 Pentaerythritol as a candidate in the recycling of polyurethane foam wastes under microwave irradiation

3.8.1 Introduction and background

Chemical recycling of waste polymers (especially condensed polymers) has been postulated by some researchers to be the main aim of recovering valuable products. Among chemical-recycling processes, hydrolysis and glycolysis have been the most popular because they can convert PU wastes to starting materials (polyols) which can be used to produce medium-quality products or be blended with raw materials. In the glycolysis process, use of an excess amount of glycolysis agent (much greater than the stoichiometric quantity) enables the reaction product to separate into two phases, in which the upper-phase is mainly the recovered polyol. The main goal of this research was introduction of a new green destroying reagent using a mixture [glycerol in combination with pentaerythritol (PER)] and MW irradiation (as a rapid heating source) to obtain an amine-free polyol. In this work, we investigated a simple and environmentally friendly method using glycerin/PER green media and MW irradiation to recover a polyol from PUF [9].

3.8.2 Experimental section

3.8.2.1 Materials and methods

The PUF was obtained from Daltoflex® EC 20,240 (formulated virgin polyol, 100 pbw) and SUPRASEC® 2027 (MDI, 65 pbw). A recommended mould temperature of 40–45 °C was employed with a demoulding time of 5 min and a targeted moulding density of 42–47 kg/m^3. In all reactions, split-phases were obtained. The upper-phase was characterised as a polyol and the lower-phase was an amine-rich product that was applicable as a portion of rigid polyol in formulations of a new rigid PUF. Samples of rigid PUF were formulated using virgin (and blends with recycled) polyol (DALTOFOAM® TA 14,066) at 100 pbw, and SUPRASEC® 2027 at 133 pbw. PER was produced by DSM Nutritional Limited (Canada) at 99.94% assay and contained 0.61% water.

3.8.2.2 General procedure

Scraps of flexible PUF (5 g) were segmented into small sizes. Pieces were placed together with NaOH (1% *w/w*) and glycolysis agents in a three-necked flask, and then placed in a MW oven at 180 °C and 800 W for different reaction times (Table 3.19). In all reactions, the ratio of degradation agent to foam was 1:1, and the dissolution time was measured at 180 °C and 800 W, and set as the end of reaction time. After reaction completion, the reaction mixture was removed from the MW oven and allowed to cool at RT. Suspended solids and unreacted foams were filtered, and the filtrate liquid transferred until a split into two phases was observed. Finally, the upper-phase (recovered polyol) was characterised by spectroscopic methods.

3.8.3 Results and discussion

Using a MW method for glycolysis gave the opportunity to use ≤30% PER in the solvent mixture without dissolution of the foam. Some increments in viscosity were observed (especially in the bottom- phase). The glycolysis reactions of flexible PUF were carried out at split-phase, in which the upper-phase consisted of recovered polyol and the lower-phase was composed of aromatic diamines, transesterification byproducts, excess unreacted glycerin, and PER. For comparison and evaluation, upper- and lower-phase products were collected and weighed for quantitative evaluations (Table 3.19).

Table 3.19: Different ratios of destroying agent in chemical recycling of PU.

Lower-phase (%)	Upper-phase (%)	Time (min)	PER (%)	Glycerol (%)	Sample
40	60	0.4	5	95	GPF1
30	70	0.5	10	90	GPF2
50	50	1	15	85	GPF3
40	60	2	20	80	GPF4
40	60	3	25	75	GPF5
30	70	5	30	70	GPF6

Reproduced with permission from M.M. Alavi Nikje and F. Haji Aga Mohammadi, *Polymer–Plastics Technology and Engineering*, 2010, **49**, 8, 818. ©2010, Taylor & Francis [9]

As shown in Table 3.19, reaction times increased with increasing PER content in the degradation solvent. That is, the best reaction conditions were obtained using 20% PER and by increasing PER content ≤30%. One of the most important purposes of this process was application of the lower-phase in formulations of rigid PUF. Viscosity is an important physical parameter in PUF formulations, so this phenomenon was critical for the viscosity of the lower- phase. To evaluate the effects of different PER concentrations in the composition of degradation solvent, we followed reactions for ≤30% PER. In all reactions, the recovery of polyol was observed successfully in the upper-phase, but the bottom-phase tended to yield greasy and unusable material for formulation of rigid PUF. The increasing viscosity of the lower-phase occurred due to the tendency of PER to crystallise and solidify. Figure 3.31 shows the chemistry of the reaction. The first step is nucleophilic attack of a transesterifiable reagent to the urethane bond, which is accomplished by formation of dicarbamates and polyol recovery.

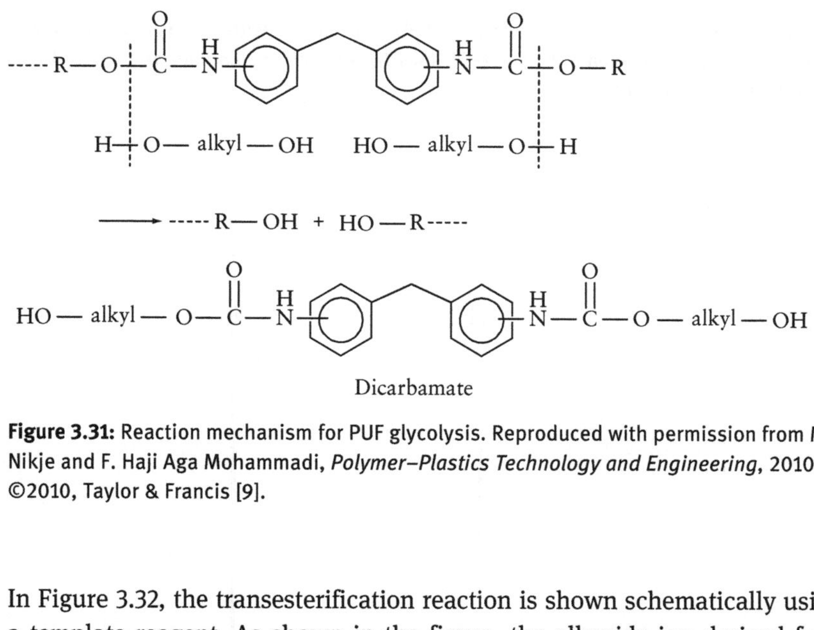

Figure 3.31: Reaction mechanism for PUF glycolysis. Reproduced with permission from M.M. Alavi Nikje and F. Haji Aga Mohammadi, *Polymer–Plastics Technology and Engineering*, 2010, **49**, 8, 818. ©2010, Taylor & Francis [9].

In Figure 3.32, the transesterification reaction is shown schematically using PER as a template reagent. As shown in the figure, the alkoxide ion derived from PER is reacted in the urethane bond, and the original polyol is recovered as the reaction product.

3.8.3.1 Characterisation of reaction products

One of the most important results in our studies was the increasing purity of the recovered polyol after additional MW irradiation. Figure 3.33 shows the ^{1}H-NMR spectra of virgin polyol a), recovered polyol b) and recovered polyol after 2 min

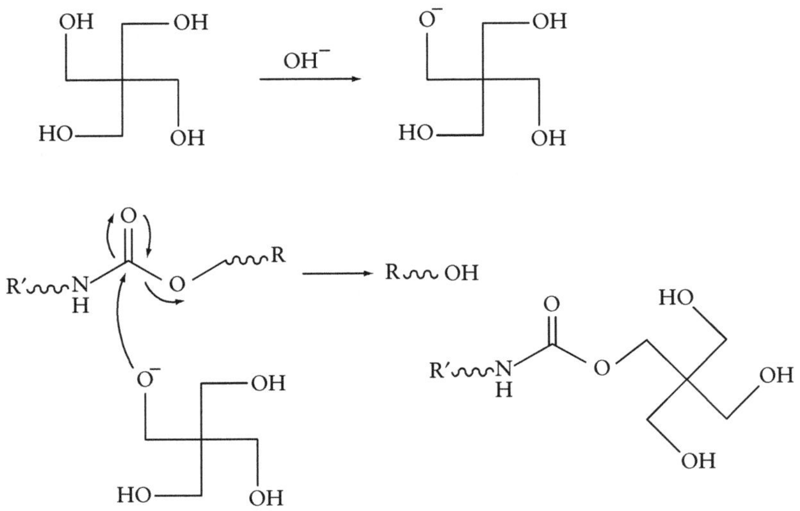

Figure 3.32: PER as a template in chemical recycling of PU. Reproduced with permission from M.M. Alavi Nikje and F. Haji Aga Mohammadi, *Polymer–Plastics Technology and Engineering*, 2010, **49**, 8, 818. ©2010, Taylor & Francis [9].

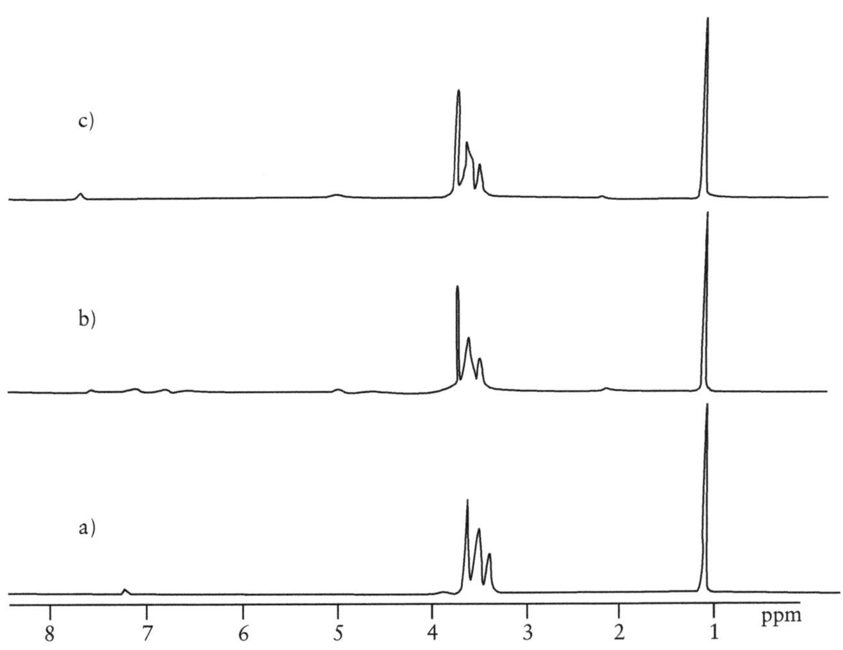

Figure 3.33: ^{1}H-NMR spectra of virgin polyol a) recycled polyol b) and recycled polyol after additional MW irradiation for 2 min c). Reproduced with permission from M.M. Alavi Nikje and F. Haji Aga Mohammadi, *Polymer–Plastics Technology and Engineering*, 2010, **49**, 8, 818. ©2010, Taylor & Francis [9].

additional MW irradiation c), respectively. According to ^{1}H-NMR assignments, the chemical structure of the recycled polyol was similar to that of the virgin material. The slightly weak signals in the 7.0–7.3 ppm region of spectra b) corresponded to residual aromatic byproducts derived from the starting polyisocyanate that were partially soluble in the recovered polyol. As shown in spectrum c), when the reaction mixture was irradiated for 2 min the weak signals disappeared and amine- free polyol was recovered without the need for further purification.

Comparison of the FTIR spectra of recovered polyols demonstrated the ability of the degradation agent to breakdown PU linkages. The FTIR spectra of virgin a) and recycled b) polyols are shown in Figure 3.34 and signal assignments are presented in Table 3.20, respectively. Distinctive absorption bands at 1,115, 3,482, 2,868, 1,456 and 1,374 cm^{-1} were observed in the FTIR spectra of the virgin polyol. Similar absorption bands were present in the FTIR spectrum of recovered polyols except for two new absorption bands at 1,627 and 1,515 cm^{-1} and also the absorption band at 1,728 cm^{-1}, which disappeared by additional MW irradiation. The FTIR spectrum of an amine-free recovered polyol is presented in Figure 3.35, which suggested excellent agreement to that obtained from the virgin polyol.

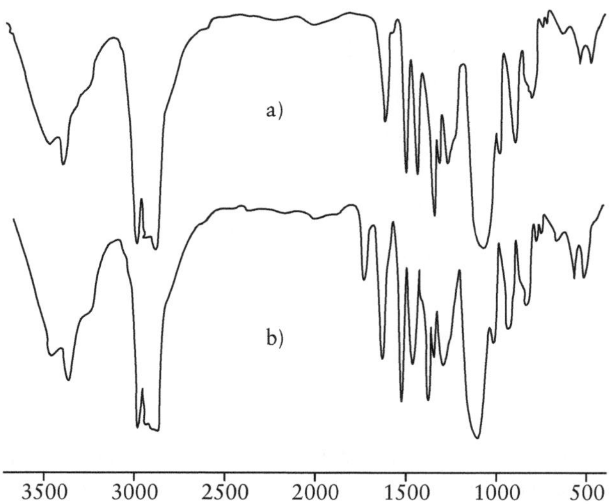

Figure 3.34: FTIR spectra of virgin a) and recycled b) polyols. Reproduced with permission from M. M. Alavi Nikje and F. Haji Aga Mohammadi, *Polymer–Plastics Technology and Engineering*, 2010, **49**, 8, 818. ©2010, Taylor & Francis [9].

3.8.3.2 Preparation of new rigid polyurethane foams

Another important observation in these reactions was the compatibility of the recycled polyol with the virgin polyol to ensure it could be used in similar

Table 3.20: FTIR spectra data comparisons of virgin and recycled polyols.

Wave number (cm^{-1})	Assignment of functional group
1,115	Stretching vibrations of ether (–C-O-C-) groups
3,482	Hydroxyl groups associated with hydroxyl end groups
2,868	Stretching vibrations of -CH bonds
1,456 and 1,374	Bending vibrations of – CH_2 and – CH_3 groups
1,728	Urethane oligomers (appear in spectrum of recycled polyol)
1,627 and 1,515	Bending vibrations of the amine group (appeared in spectrum of the recycled polyol)

Reproduced with permission from M.M. Alavi Nikje and F. Haji Aga Mohammadi, *Polymer–Plastics Technology and Engineering*, 2010, **49**, 8, 818. ©2010, Taylor & Francis [9]

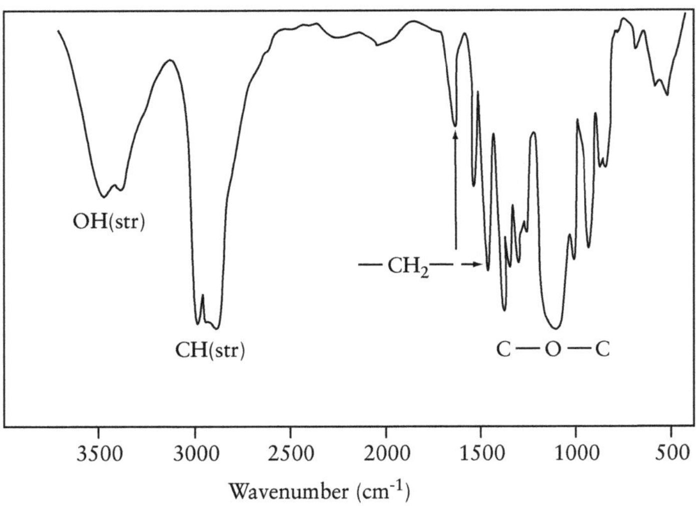

Figure 3.35: FTIR spectrum of recycled polyol after additional MW irradiation for 2 min. Reproduced with permission from M.M. Alavi Nikje and F. Haji Aga Mohammadi, *Polymer–Plastics Technology and Engineering*, 2010, **49**, 8, 818. ©2010, Taylor & Francis [9].

formulations for manufacture of new rigid PUF. We focused on reactivities, densities, and foam qualities in this research, and data are collected in Table 3.21. According to the table, increasing the lower-phase to 20% caused an increment in density and lowered the quality of new rigid PUF. These results suggested the lower-phase at ≤20% could be used in formulations of rigid PUF.

Table 3.21: Reactivity parameters, density and foam qualities of samples of rigid PUF during formation.

Foam quality	Density (g/cm^3)	Tack free (s)	String time (s)	Rising time (s)	Cream time (s)	Sample
OK	0.052	112.0	65.0	60.0	25.5	PU1
OK	0.092	40.0	60.0	55.0	31.3	PU2
OK	0.119	36.0	39.0	53.8	33.0	PU3
OK	0.125	22.1	27.0	46.0	36.0	PU4
OK	0.171	10.0	16.0	36.0	38.0	PU5
Failed	0.126	4.0	10.0	31.0	40.0	PU6
Failed	0.081	–	–	25.0	45.0	PU7

Reproduced with permission from M.M. Alavi Nikje and F. Haji Aga Mohammadi, *Polymer–Plastics Technology and Engineering*, 2010, **49**, 8, 818. ©2010, Taylor & Francis [9]

3.9 Polyvalent alcohols as the glycolysing agent in recycling of wastes of flexible polyurethane foam

3.9.1 Introduction and background

In recent years, public awareness of environmental issues has increased, especially of the facts that the world has limited natural resources and a limited capacity to manage the volume of wastes which human activities produce. Recycling of polymeric products has been studied to enhance environmental protection and avoidance of landfilling. Among various polymers, recycling of PU has always posed some challenges. Chemical recycling of PU by hydrolysis or aminolysis is, for the most part, considered to be economically uncompetitive compared with formulation with virgin raw materials. The high temperature and pressure required for hydrolytic decomposition of urethane structures, in addition with difficulties in reformulation of aminolysis products, have prevented these processes being feasible or used at the industrial scale.

The glycolysis process is the most widely used chemical-recycling method for PU wastes. The main purpose of glycolysis is recovery of polyols from waste PU for the manufacture of new materials. A method for preparing a recyclate polyol obtained from glycolysis of PU comprises reacting scrap PU with a short-chain compound containing ≥2 hydroxyl groups in the presence of a catalyst with addition of a cyclic carbonate. The process results in a recyclate polyol having a low content of primary aromatic amines that may be used to manufacture cellular or non-cellular PU, polyurea-PU, or polyisocyanurates. The reacting agent is typically DEG and co- reagent is DEA. A reaction temperature <180 °C lowers catalyst activity severely and >220 °C produces undesired side reactions towards amines. With regard to the catalyst, formation of aromatic amines must be avoided.

In this research, an attempt was made to use DEG–PER as a new solvent–reactant mixture to evaluate recyclate specifications and also the effect of recyclate on formulated foams. PER contains 4 hydroxyl functional groups, which would accelerate the dissociation of urethane structures. Its multi-active sites for conducting a recycling method and its similarity with polyol structures are highly important. The upper-phase of the recyclate is useful for the manufacture of flexible PUF. Also, regeneration of polyol by PER-assisted recycling of PU provides a high ratio of OH/Mw in the lower-phase, which is useful for the manufacture of high-quality rigid PUF [10].

3.9.2 Experimental section

3.9.2.1 Materials and instruments

DEG, PER and NaOH were used. Polyether polyols (BAYNAT® PU-80IF04 for flexible foam and DALTOFOAM® TA 14,066 for rigid foam) containing all the required additives, and isocyanate (MDI-SUPRASEC® 5005) were used for PUF formulation. Spectroscopy-grade THF (Merck) and polyethylene glycol standards (Merck) for GPC analyses were used. The M_WD was measured by GPC. Chromatograms were obtained using a 6A GPC instrument (Shimadzu). Water content was determined by a D18 Karl Fischer auto-titrator (Mettler). Tensile strength of PUF samples was measured according to ASTM D638 using an 11-model tensometer (Instron).

3.9.2.2 Degradation of flexible polyurethane foam in a diethylene glycol–pentaerythritol mixture

Flexible PUF was decomposed by a mixture of DEG and PER (in different mass ratios) in the presence of NaOH as the catalyst. A reflux system containing a 1-l three-necked, round-bottom flask equipped with a condenser, thermometer, and agitator was used. Solvent mixture (200 g) was placed in the flask and heated to 160 °C. Then, 1 g NaOH was dissolved in the mixture, temperature was increased to 200 °C, and 200 g of shredded PUF added to the flask (15 g/min). Digestion occurred as the foam came into contact with the solvent at an acceptable rate. Degradation was continued for 5 h at atmospheric pressure. The reaction product was separated into two phases of (upper) clear-yellow solution and (lower) light-brown solution upon cooling after 24 h. Both phases of the product were characterised using instrumental and classical analytical methods.

3.9.2.3 Preparation of samples of polyurethane foam by product phases

According to characterisation of the recycled product, the upper- and lower-phases were used in the formulations of flexible and rigid PUF, respectively. Phases were separated and the one-shot method was applied for foam production.

Polyol blends and additives were premixed in a 300-ml carton beaker by 2-min mixing followed by MDI addition. The pre-polymer was mixed for 10 s (beyond the creaming time) and then allowed to rise freely, remaining at RT for the post-cure step in the next 24 h. Stoichiometry of the starting formulation for flexible PUF was 100 pbw of BAYNAT® PU 80IF04 and 65 pbw of SUPRASEC® 5005. The virgin rigid PUF was prepared usinf 100 pbw of DALTOFOAM® TA 14,066 and 133 pbw of SUPRASEC® 5005. The virgin polyol was substituted with the recycled product at various ratios.

3.9.3 Results and discussion

3.9.3.1 Reaction set

Previous studies have confirmed that cleavage of urethane bonds occur in the presence of an alkaline metal hydroxide as a catalyst. A set of reactions was undertaken using various ratios of DEG–PER. PER contains 4 hydroxyl groups, so has high functionality for transesterification during glycolysis. Unreacted PER was vacuum-filtered after phase separation. In all reactions, 100 g of PUF was reacted. A high amount of PER reduces the reactivity of hydroxyl groups in the solvent systems. Interestingly, the product was increased by increasing the DEG:PER ratio, whereas that of the evaporated mixture was lowered. According to the data obtained in our experiments, the DEG:PER ratio was set as 9:1, and it was under this condition that other experimental parameters were studied.

To optimise the reaction, two parameters were studied: the Mw of product-phases by GPC and the remaining DEG in recycling product by high-performance liquid chromatography (HPLC). To determine the Mw of products, PEG of various Mw (400, 800, 1,000, 1,600, 2,000, 4,000 and 10,000) were employed as internal standards to create the calibration model.

HPLC chromatograms of the upper-phase in the recycling product showed that the DEG content of both phases was near to 0.0%. This was an important parameter needed to obtain the best reaction time, whereby the Mw of both phases would be at a minimum for time- specified reactions. The obtained product mixtures from reaction 6 have been separated to split-phases when cooled overnight and left for 4 h.

3.9.3.2 Fourier-transform infrared analyses of regenerated polyols

FTIR spectroscopy was used to investigate the chemical structure of recycled products and compared them with virgin polyols. Some of the main similarities, observed in the molecular structure of recycling products and virgin polyols, are explained below:

- BAYNAT® PU 80IF04 and upper-phase:
  - Intensive band of the aliphatic ether group related to hydroxyl end-groups at 1,110 cm^{-1}.

  - Bending vibrations of methylene groups at 1,372 and 1,456 cm^{-1}.
  - Stretching vibrations of carbon–hydrogen bonds in aliphatic carbons at 2,870–2,971 cm^{-1}.
  - Stretching vibration of hydroxyl group at 3,486 cm^{-1}.
- DALTOFOAM® TA 14,066 and lower-phase:
  - Stretching vibrations of hydroxyl groups and C-O ether at 1,105 cm^{-1}.
  - Stretching vibrations of methylene group at 2,871–2,970 cm^{-1}.
  - Bending vibrations of methylene groups at 1,375, 1,465, 1,520 and 1,625 cm^{-1} and also assigned to remaining additives.

3.9.3.3 Comparing regenerated polyols with virgin polyols

Specific gravity (in g/cm^3) of separated phases was similar to those of virgin polyols (flexible polyol: 1.03; upper-phase: 1.09; rigid polyol: 1.06; lower-phase: 1.13). One of the most important factors affecting injection in PU machines is polyol viscosity. Higher viscosities cause several problems (e.g., heterogeneity in prepolymers, overloaded pressure of polyol stream which destroys the injection moulding machine). Viscosity of BAYNAT® PU 80IF04 was 1,740 cps, whereas that of the upper-phase of the recycled polyol was 1,592 cps. Regenerated polyol in the lower-phase had a viscosity of ≈5,660 cps, whereas that of the virgin rigid polyol was 5,260 cps (all at 25 °C).

Water as a blowing agent has an important function in PUF formation. Water content of both virgin polyols and both regenerated polyols were measured using an automated Karl Fischer titrator. Water contents (in *w/w%*) of both product-phases were less than those of virgin polyols (flexible polyol: 1.2; top-phase: 0.8; rigid polyol: 2.3; lower-phase: 1.3 *w/w%*). The pH of all four polyols was measured using a digital pH meter. The pH was similar for the flexible polyol and upper-phase (10.1 and 9.8, respectively) but, with higher amounts of alkaline components in the lower-phase of the recycled product, the pH was higher in comparison with the virgin rigid polyol (9.9 and 9.5, respectively). The most important factor is the Mw, which determines the application potential of products-phases. A higher Mw of hydroxyl-containing products provides higher crosslinking density and flexibility of the formulated foam. Upper- and lower-phases were applied in cold-cure flexible and rigid foam formulations, respectively.

3.9.3.4 Application of product-phases in new foam formulations

The main aim of this research was to regenerate raw materials and, according to comparative tests on the product, the upper-phase was reused for production of flexible foam and the lower-phase was used in production of rigid foam. The starting formulation of typical cold-cure flexible and rigid PUF is given in detail in the experimental section. It is evident that >30% of the recycled polyol can be tolerated by increasing MDI use. The narrow M_WD of the recycled polyol enhanced the functionality that parallels the hydroxyl-value of the polyol. One of the other methods to compare recycled and virgin polyols is to compare their reactivities by recording the time factor for foaming.

Regenerated phases were used in polyol blends at ≤50%. All ingredients and their quantities in the synthetic process of foam, in addition to reactivity parameters (time control), were studied in detail.

In test measurements for physical properties, two regenerated foams were compared with those of virgin-based foams. Physical properties of foam samples containing regenerated polyols suggested reliable results which would ensure application of recycling products in foam formation. That is, improvement in the physical properties of rigid PUF was in accordance with incremental increases in the crosslink density of the produced foam in the presence of PER-based oligomers. These oligomers were hyperbranched and highly functional. Some characteristics of flexible PUF were decreased in comparison with open-cells, which are more prominent in recycled polyol-containing samples because of the high functionality of low-Mw oligomers.

3.10 Polyurethane waste and starch liquefaction products as polyols for formulation of polyurethane foams

3.10.1 Introduction and background

PUF are often used as polymeric materials and, due to their versatile characteristics, production of a great volume of waste is inevitable. Also, PU are non-biodegradable materials and formation of high- volume waste during production as well as aged and used materials leads to environmental damage. One of the most important processes used to overcome environmental drawbacks is recycling, and it is recommended for recovery and regeneration of valuable materials from waste. Among recycling processes, the chemical method is versatile and economical. Chemical recycling and synthesis of biodegradable PUF can be employed as versatile methods to reduce waste. Designing a combined process can bring new opportunities in reducing waste of PU and other polymeric waste. By applying MW irradiation as an energy source for carrying out chemical reactions, a considerable reduction in reaction times is observed. Using MW irradiation produces a versatile and economic method.

In this research, starch-based PUF were obtained in the reaction of a polyol blend derived from MW-assisted liquefied corn starch product and chemically recycled polyol from cold-cured PUF waste, followed by blending with a virgin polyol and MDI as the isocyanate portion. The effect of MW irradiation programmes was studied in starch liquefaction and recycling of PUF waste to ascertain optimum conditions. Reactivity factors, densities and thermal properties of synthesised starch/recycled polyol-based foams were investigated [11].

3.10.2 Experimental section

3.10.2.1 Materials and instruments

PUF for chemical recycling was formulated using flexible foam-grade polyol (Daltoflex® EC 20,240 and MDI-SUPRASEC® 5005). Virgin polyol used for preparation of rigid PUF was DALTOFOAM® TA 14,066. Starch was purchased from a corn starch-producing company in Iran and used without further treatment. Recycling of PUF and polyol recovery was done using methods described above. The recovered polyol was characterised by spectroscopy methods.

3.10.2.2 Microwave-assisted reactions

MW-assisted reactions were done in a MicroSYNTH 'NP' Ethos 1600 MW oven (Milestone) at atmospheric pressure.

3.10.2.3 General procedure for microwave-assisted recycling of polyurethane foam

For recycling, scraps of flexible PUF (5 g) (segmented into small sizes), NaOH (1% *w/w*) in the presence of solvent mixtures comprising glycerol (95%), sorbitol (3.5%) and water (1.5%) were poured into a two-necked flask. The mixture was placed into a MW oven under atmospheric pressure and irradiated at 180 °C and 800 W. When foam digestion was completed, the reaction mixture was removed from the MW oven and left to separate into two distinguishable phases. Obtained polyols from the upper-phase were decanted, centrifuged for 20 min, and characterised by spectroscopic methods.

In all the recycling procedures studied here, split-phases were formed. The upper-phase was employed as a portion of polyol in the PUF formulation.

The aim of this research was evaluation of efficiency of the recycled polyol in new starch-based PU formulations. Hence, the upper-phase was used directly in combination with other components and as the polyol component for formulation of rigid PUF. The recycled polyol was first identified structurally using ^{1}H-NMR, ^{13}C-NMR and FTIR and data compared with an authentic sample and our previous results.

Results from OH# determination of the recycled polyol demonstrated that it had a higher value than that of the virgin polyol (59 mg KOH g^{-1}). This phenomenon was due to the partial solubility of glycerine and/or sorbitol in the upper-phase, which is important for increasing the hydroxyl content of the recycled polyol. Hydroxyl values for recycled polyols were determined to be 73 and 64 before and after centrifugation, respectively. Another important property that affects the final properties of the foam directly is the water content of the recycled polyol, which was determined to be 1.80 ± 0.1%. Determination of water content in the recycled polyol was less than that of the virgin polyol, which was applied for manufacture of final rigid PUF.

3.10.2.4 General procedure for microwave-assisted starch liquefaction

Green solvents (DEG and glycerin and/or blends thereof), catalyst and powdered starch mixtures in different ratios were irradiated in a MW oven under controlled conditions.

After completion of the liquefaction reaction, the OH# of liquefied starch was determined to be 430 ± 20 and water content to be 0.55%, respectively.

Thermogravimetric analysis (TGA) was carried out with a Pyris Diamond TG/DTA system (Perkin–Elmer) at a heating rate of 20 °C/min under a N_2 atmosphere. Foam samples were prepared in a paper cup by hand mixing.

3.10.3 Results and discussions

3.10.3.1 Microwave irradiation programme for starch liquefaction

DEG and glycerin and mixtures thereof were employed as liquefaction solvents. Reactions were investigated under acid-catalysed and uncatalysed conditions. To investigate the roles of solvents in starch liquefaction, all experiments were done under controlled MW conditions, as noted above.

Power (≤500 W) was at 3 min and the bulk temperature was held at 120 °C. Liquefaction continued by decreasing the MW power to 300 W in the second step and holding the bulk temperature fixed at 120 °C. Liquefaction was monitored visually and formation of a clear solution was the criterion for completion of the process.

Obtained results showed unsuccessful liquefactions in uncatalysed reactions. Data suggested that performance of the process in acidic media broke down starch molecules and converted high-Mw starch polymers to the corresponding oligomers. In uncatalysed reactions, dispersion of starch molecules occurred in solvent media and starch particles were precipitated during cooling.

Performances of an acid catalyst and MW irradiation have considerable effects upon liquefaction, as well as upon chain breaking of starch. Results showed that, under intensive conditions, reactions that break down chains are combined with excessive oxidation reactions. That is, liquefaction carried out without an excessive oxidation reaction was successful. Among the successful liquefaction reactions, one of them considered as the convenient condition in this research due to formation of a stable media and reaching the viscosity to the virgin polyol viscosity and further studies has been performed on this solution.

3.10.3.2 Starch-based preparation of polyurethane foams and determination of reactivity factors

All samples were prepared by hand mixing according to the specified weight ratios. A cup test was carried out to determine reactivity factors. The latter have significant effects upon foam production and affect operational and RIM conditions as well as the final foam properties, so we focused our efforts on determination of these factors.

In the experiments, the starch solution ratios were considered to be constant in all formulations and the portion of the recycled polyol was varied slightly to study reactivity factors. By increasing the amount of recycled polyol, reaction times increased due to decreasing concentrations of additives in the formulated virgin polyol. All samples were stable in dimension after 24 h of foam formation without formation of holes or shrinkage. During foam formations, destructive exothermic reactions were not observed, which revealed a starch-based polyol in the new PUF formulation.

3.10.3.3 Density determination

Density tests were carried out for all sample foams by cutting samples from within the foams. Data from this experiment revealed that densities were increased by increasing the content of recycled polyol in foam formulations. Due to the low solubility of destroying agents in the upper-phase, a higher OH# value was expected in comparison with that of the virgin polyol. Formulated foams were denser when the recycled polyol content in the formulation was increased, and the formulated foams were denser in comparison with foams obtained with liquefied starch/virgin polyol in a blend with small recycled polyol portions.

3.10.3.4 Thermal gravimetric analysis studies

Thermal behaviours of prepared foams were investigated to ascertain the effects of starch and recycled polyol in the thermal resistivity of foams. A blank sample PUF formulated by wholly commercially available polyol and isocyanate was considered, and other foams data compared with blank sample data. Decomposition temperature of 10% of sample mass (T_d), 10% and T_d, 80% was assigned as the target 10% and 80 wt% temperatures, respectively. From TGA, the highest thermal resistance was attributed to a sample containing 45% of recycled polyol. By increasing the content of the recycled polyol to >45%, the foam collapsed and foaming was incomplete. This phenomenon was expected to be due to insufficient foaming-agent content in the sample, which decreased in concentration with increasing content of recycled polyol.

References

1. M.M. Alavi Nikje, M. Haghshenas and A. Bagheri Garmarudi, *Polymer–Plastics Technology and Engineering*, 2006, **45**, 4,569.
2. M.M. Alavi Nikje, M. Haghshenas and A. Bagheri Garmarudi, *Polymer Bulletin*, 2006, **56**, 2, 257.
3. M.M. Alavi Nikje, M. Haghshenas and A. Bagheri Garmarudi, *Polymer–Plastics Technology and Engineering*, 2007, **46**, 3, 265.

4. M.M. Alavi Nikje and M. Nikrah, *Polymer Bulletin*, 2007, **58**, 2, 411.
5. M.M. Alavi Nikje, M. Nikrah and M. Haghshenas, *Polymer Bulletin*, 2007, **59**, 1, 91.
6. M.M. Alavi Nikje and M. Nikrah, *Polymer–Plastics Technology and Engineering*, 2007, **46**, 4, 409.
7. M.M. Alavi Nikje, M. Nikrah and F. Haji Aga Mohammadi, *Journal of Cellular Plastics*, 2008, **44**, 5, 367.
8. M.M. Alavi Nikje and F.Haji Aga Mohammadi, *Polimery Journal*, 2009, **54**, 7–8, 541.
9. M.M. Alavi Nikje and F. Haji Aga Mohammadi, *Polymer– Plastics Technology and Engineering*, 2010, **49**, 8, 818.
10. M.M. Alavi Nikje and A. Bagheri Garmarudi, *Iranian Polymer Journal*, 2010, **19**, 4, 287.
11. M.M. Alavi Nikje, *Polimery*, 2012, **57**, 1, 11.

4 Chemical recycling of rigid polyurethane foams via thermal and microwave-assisted methods

4.1 Starch, sucrose and Rezol® IL800 as density modifiers for polyurethane rigid foams formulated by recycled polyols

4.1.1 Introduction and background

Rigid polyurethane foam(s) (PUF) are heat-insulating materials. They are one of the principal components of home appliances or heat-insulation systems. The furniture industry, building construction, heat-insulation systems, transportation (including automotives), the shoe industry, and aviation are some of the many PUF consumers. Rigid PUF have low density (25–150 kg/m^3) most of the time, so few mass of them occupy large spaces, which leads to environmental pollution because of loss of these polymers. In this research, we first recycled rigid PUF using glycolysis (according to condition detailed later), formulated a new rigid PUF, and then identified some additives to modify the density of rigid PUF formed by application of a recycled polyol [1].

4.1.2 Experimental section

4.1.2.1 Materials and instruments

The waste rigid PUF used for recycling was from refrigerator and pipe-insulation wastes that had a light-yellow colour and which were crushed into pieces of diameter 1 cm. Diethylene glycol (DEG), diethanolamine (DEA) and sodium hydroxide (NaOH) were obtained from commercial sources. The virgin polyol (DALTOFOAM® TA 14,066) and 4,4′-diphenylmethane diisocyanate (MDI) (SUPRASEC® 5005) were used for PUF manufacture. The prefix 'virgin' was added to the main polyol (non-recycled polyol) to distinguish it from the recycled polyol. The virgin polyol was provided as a system containing all additives required for foaming. The recycled polyol was selected from the product of a glycolysis-recycling process of rigid PUF that was formulated with the same raw materials. Starch and sucrose were of commercial grade, and REZOL® IL800 is a commercial product of the Rezitan Company in Iran (Table 4.1). Water contents were determined using the Karl Fischer method, and viscosities were measured using a viscometer (Haake). pH values were determined using a pH meter (Schott Gerate). A 6A gel permeation chromatography (GPC) instrument (Shimidzo) was used to obtain the molecular weight (Mw) range and its abundance in polyols.

https://doi.org/10.1515/9783110641592-004

Table 4.1: Technical data for Rezol® IL800.

Appearance	Reddish-brown liquid
Viscosity	600–800 cps at 20 °C
Specific gravity	1.21 g/cm^3 at 20 °C
Effective life	Maximum 4 weeks
Maintenance condition	At 20 °C in closed tanks

Reproduced with permission from M.M. Alavi Nikje and A. Bagheri Garmarudi, *Polymer–Plastics Technology and Engineering*, 2006, **45**, 10, 1101. ©2006, Taylor & Francis [1]

4.1.2.2 Recycling of rigid polyurethane foams

In the recycling process, a reflux system containing a 1-l round-bottom flask equipped with a condenser, thermometer, and agitator was applied. Glycolysis was carried out at atmospheric pressure. Two-hundred and five grams of the solvent system (containing 190 g DEG, 10 g DEA, and 5 g NaOH) were placed in the reactor first. The speed of the stirrer was set at 700 rpm and the temperature of the heating mantel was set at 200–205 °C. When the temperature of the mixture (DEG/DEA/NaOH) reached the specified value, 410 g of crushed rigid PUF was poured into the reactor at a feeding rate of 4–5 g/min. Glycolysis was continued up to 5.5 h of reaction time. The product's properties were determined using instrumental and classical methods (Table 4.2).

Table 4.2: Technical data of recycled polyols used in PUF formulations.

Appearance	Viscous light-brown liquid
Viscosity	5,800 cps at 25 °C
Specific gravity	1.11 g/cm3 at 25 °C
Water content	0.4%
pH	8.8 ± 0.1
OH#	450 ± 20

OH#: Hydroxyl number
Reproduced with permission from M.M. Alavi Nikje and A. Bagheri Garmarudi, *Polymer–Plastics Technology and Engineering*, 2006, **45**, 10, 1101. ©2006, Taylor & Francis [1]

4.1.2.3 Preparation of foams

Foams were obtained using the one-shot method. Ingredients were mixed together at once; MDI was added and allowed to cure. Polyol blend, modifier additive, and water (if needed) were mixed in a 300-ml carton beaker for 2 min by mechanical stirring followed by addition of MDI to the mixture and stirring for 10 s until whitening of the liquid ('cream time'). Samples were allowed to remain at room temperature (RT) (20 ± 2 °C) for complete curing during the next 24 h.

The stoichiometric proportion of raw materials was 1:1 (polyol:MDI), which gave a foam with a density of 34–37 kg/m^3 at room temperature (RT).

4.1.2.4 Typical process

The aim of this research was to ascertain the effect of starch, sucrose and Resol® on the density of rigid PUF. Attending to the physical appearance of foams was needed to access results applicable for industrial formulation of foams. First, each of additives was added to the foam formulation, and then density measurement was carried out on each sample according to the American Society for Testing and Materials (ASTM D1622). Results were compared to ascertain the best formulation. The foam formulated using a 30% recycled polyol blend had a density of 49.25, but decreased to 43.93 kg/m^3 when extra water (≤2.3%) was added to the sample.

4.1.3 Results and discussions

4.1.3.1 Characteristics determination of recycled polyol

Recycled polyol was manufactured as an unknown material, so it was important to determine the specifications that may influence foaming. Hence, several instrumental and classical analytical methods were carried out on it (some on both polyols) to achieve the necessary information:

- *GPC analyses*: GPC was used to analyse the distribution of Mw. Results indicated that the main Mw range was 300–800 for a virgin polyol having 76.19% of abundance for the virgin polyol and 60.01% abundance for the recycled sample.
- OH#: Determination of the hydroxyl value for virgin and recycled samples was done according to ASTM D4274, and was 430 ± 20 and 450 ± 20 mg of potassium hydroxide (KOH) for virgin and recycled polyol, respectively. The theoretical OH# was calculated according to the main Mw and its abundance which, according to GPC analyses, was 330 mg of KOH for the virgin polyol and 350 mg of KOH for the recycled polyol.
- *Viscosity optimisation*: As mentioned previously, the viscosity of recycled polyol was 5,800 cps at 25 °C. Reaction injection moulding (RIM) was carried out at 40–60 °C. The viscosity of the recycled polyol was 1464, 670.2 and 338.3 cps at 40, 50 and 60 °C, respectively. Also, the viscosity of a blend containing 30% recycled polyol was 5,419 cps.

- *Physical properties: comparison of foams formulated by virgin and recycled polyols*: In the next step, we tried to compare the physical properties of two sample foams: one containing 30% recycled polyol in its polyol blend and the other formulated just by the virgin polyol. The MDI portion and water content in both samples were identical. Time data obtained from these two foam sample formulations during cup testing by hand mixing in the laboratory was compared (Table 4.3). Then, quality-control tests were executed on both samples (Table 4.4). As shown in Table 4.4, comparison of the physical properties of foams formulated by virgin and recycled polyols confirmed that the product of the glycolysis method was similar with the foam formulated by the virgin polyol. The product of 30% recycled polyol was low when stabilised against elongation, but it was more stable with regard to tensile strength and its elastic modulus was increased compared with that of the product from the virgin polyol. This effect was desirable because the OH# of recycled polyol was more than that of the virgin polyol. Hence, the rigidity in the product of recycled polyol was more than that of the virgin polyol. However, the density of recycled polyol-applied foam was more than that of the virgin polyol foam. We tried to regulate this disadvantage in subsequent procedures, with additives playing the part of density modifiers.

Table 4.3: Time control of PUF samples during foam formation.

Factor	Sample foam	
	PUF_1*	PUF_2**
Cream time (s)	5	5
Gel time (s)	7	8
Tack free (s)	28	25
Gas release (s)	–	–
End of rise (s)	41	39

* PUF1: formulated virgin polyol.
** PUF2: formulated by 30% recycled polyol blend.
Reproduced with permission from M.M. Alavi Nikje and A. Bagheri Garmarudi, *Polymer–Plastics Technology and Engineering*, 2006, **45**, 10, 1101. ©2006, Taylor & Francis [1]

4.1.3.2 Starch as an additive

The first step was to add starch to the formulation system just as a co-blowing agent without affecting its reaction with MDI. Enriching the polyol blend's water content to the same as that of the virgin polyol was important. Therefore, water was added at ≤2.3%. The density of samples containing starch as the only blowing agent

Table 4.4: Comparison of the quality-control results of foams.

Rigid foam properties	Virgin polyol	30% Recycled polyol
Core density (kg/m3)	34	44
Tensile strength at vertically (psi)	16.1	31.8
Elongation at break (%) vertically	10	3.5
E Modulus (psi) vertically	226	626
Tensile strength horizontally (psi)	15.7	29
Elongation at break (%)horizontally	9.2	6.25
E Modulus (psi) horizontally	201.4	560
Compression set (%)	14	11.5
Water absorption (%) (24 h)	1.5	1.2
Dimensional stability (–30 °C/24 h) (%)	OK	OK
Dimensional stability (70 °C/24 h) (%)	OK	OK

Reproduced with permission from M.M. Alavi Nikje and A. Bagheri Garmarudi, *Polymer–Plastics Technology and Engineering*, 2006, **45**, 10, 1101. ©2006, Taylor & Francis [1]

and samples containing starch and water are detailed in Table 4.5 and results are compared in Figure 4.1. In all samples, a mixture of 7 g DALTOFOAM® TA 14,066 and 3 g of recycled polyol containing the polyol blend and 10 g SUPRASEC® 5005 was added later as the MDI portion.

Table 4.5: Density of samples containing starch as an additive (blowing agent).

Starch (g)	No extra water (density kg/m)	With extra water* (density kg/m)
0	49.25	43.93
0.5	54.31	46.25
1.0	52.01	43.65
1.5	50.83	40.77
2	48.24	37.03
2.5	Foam collapsed	–

* Water was added as blend's water content reaches 2.3%. Reproduced with permission from M.M. Alavi Nikje and A. Bagheri Garmarudi, *Polymer–Plastics Technology and Engineering*, 2006, **45**, 10, 1101. ©2006, Taylor & Francis [1]

4.1.3.3 Starch as a part of a hydroxyl-containing raw material

Starch has hydroxyl functional groups and reacts with NCO groups. Thus, in the next procedure, starch was added to the system as a portion of the polyol blend,

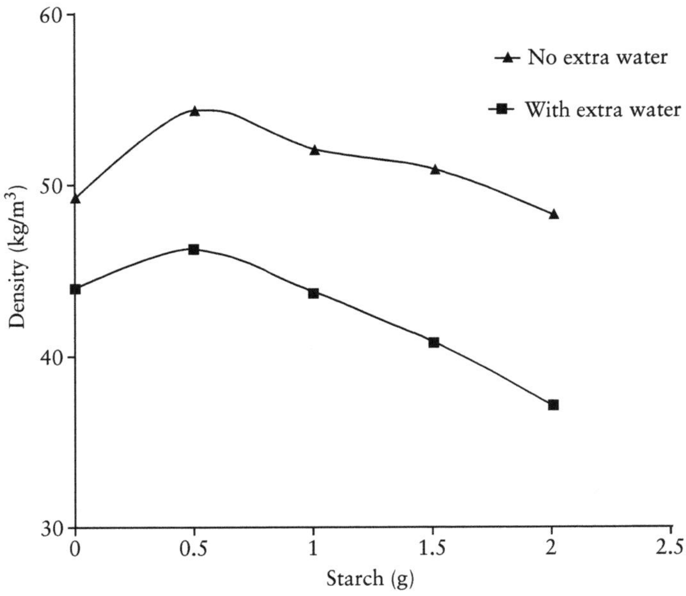

Figure 4.1: Effect of starch-blended polyol on foam density. Reproduced with permission from M.M. Alavi Nikje and A. Bagheri Garmarudi, *Polymer–Plastics Technology and Engineering*, 2006, **45**, 10, 1101.©2006, Taylor & Francis [1].

and we added more and less MDI than the start formulation to clarify the effect of the starch reaction on density. Again, the lack of water was rectified in similar samples, and density results are listed in Table 4.6 and Figure 4.2.

Table 4.6: Density of samples containing starch as a part of a polyol blend.

Virgin polyol (g)	Recycled polyol (g)	Starch (g)	MDI(g)	Water (g)	Density (kg/m^3)
6.5	3	0.5	10	0.00	40.21
6.5	3	0.5	10	0.06	33.83
6.5	3	0.5	9	0.06	30.28
6.5	3	0.5	11	0.06	35.07
6.0	3	1.0	10	0.00	43.21
6.0	3	1.0	10	0.06	37.91
6.0	3	1.0	8	0.06	32.03
6.0	3	1.0	12	0.06	36.45

Reproduced with permission from M.M. Alavi Nikje and A. Bagheri Garmarudi, *Polymer–Plastics Technology and Engineering*, 2006, **45**, 10, 1101. ©2006, Taylor & Francis [1]

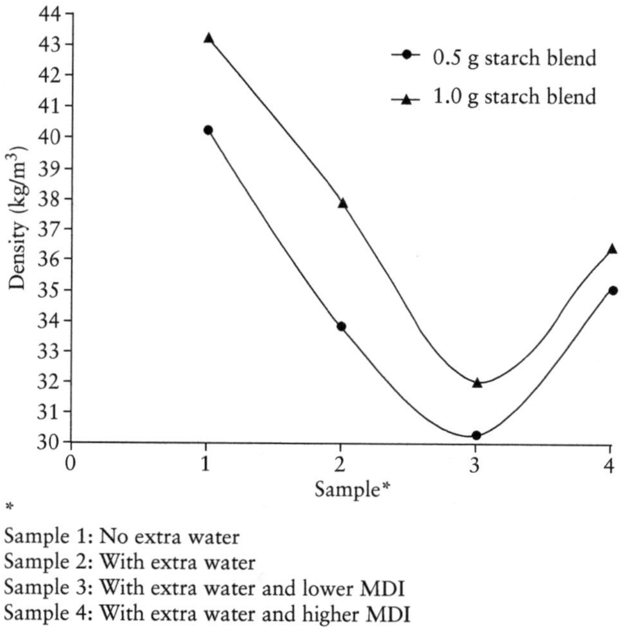

Figure 4.2: Effect of starch-blended polyol on foam density. Reproduced with permission from M.M. Alavi Nikje and A. Bagheri Garmarudi, *Polymer–Plastics Technology and Engineering*, 2006, **45**, 10, 1101.©2006, Taylor & Francis [1].

4.1.3.4 Use of the result of starch application for sucrose

All of the procedures carried out on starch suggested that the reaction between the hydroxyl groups of starch and MDI influenced foam density, and the effect of the MDI portion on foam density was defined. Therefore, in all samples formulated by sucrose, the effect of the reaction was as desired, and trying to determine the effect of MDI and water content gave us results that are compared in Figure 4.3. Proportions and densities are listed in Table 4.7.

4.1.3.5 Modification of foam density by applying Rezol® IL800

New methods became available during studies carried out on modifiers. However, it was found that the best scenario was that all parameters affecting results should be used in calculations for foam formulations together. Therefore, for Rezol® IL800, all of the previous results were used in a formulation leading to the short procedure detailed in Table 4.8. Rezol® IL800 has sufficient water content (≈3%), so extra water was not added. Another result was that Rezol® IL800 could not be added to >15% of the polyol blend because it made the foam collapse. However, this limited portion provided the requisite density. Comparison of results is shown in Figure 4.4.

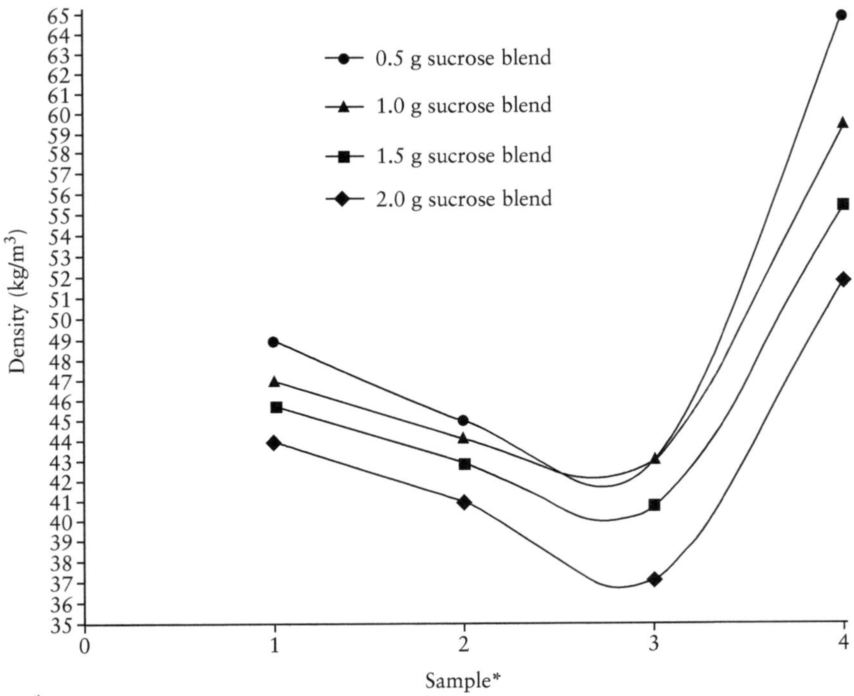

*
Sample 1: No extra water
Sample 2: With extra water
Sample 3: With extra water and lower MDI
Sample 4: With extra water and hight MDI

Figure 4.3: Effect of sucrose-blended polyol on foam density. Reproduced with permission from M.M. Alavi Nikje and A. Bagheri Garmarudi, *Polymer–Plastics Technology and Engineering*, 2006, **45**, 10, 1101.©2006, Taylor &Francis [1].

4.2 Chemical recycling and liquefaction of rigid polyurethane foam wastes through microwave-assisted glycolysis

4.2.1 Introduction and background

Rigid PUF constitutes an important group of PUF, and its demand is increasing rapidly because of broad applications such as: heat insulators in refrigerators, buildings and construction materials; floatation and packaging/padding materials; encapsulates to isolate and support thermally sensitive components within weapon systems; shock-absorbers in industry. Because of this wide range of applications, a huge volume of waste is manufactured, causing significant disposal problems. Glycolysis involves heating

Table 4.7: Effect of sucrose, MDI portion and water on the density of rigid PUF.

Virgin polyol (g)	Recycled polyol (g)	Sucrose (g)	MDI (g)	Water (g)	Density (kg/m^3)
6.5	3	0.5	10	0.00	48.85
6.5	3	0.5	10	0.06	44.93
6.5	3	0.5	9.5	0.06	42.88
6.5	3	0.5	10.5	0.06	64.66
6.0	3	1.0	10	0.00	46.91
6.0	3	1.0	10	0.06	44.02
6.0	3	1.0	9	0.06	43.11
6.0	3	1.0	11	0.06	59.34
5.5	3	1.5	10	0.00	45.55
5.5	3	1.5	10	0.06	42.83
5.5	3	1.5	8.5	0.06	40.66
5.5	3	1.5	11.5	0.06	55.37
5.0	3	2.0	10	0.00	43.88
5.0	3	2.0	10	0.06	40.91
5.0	3	2.0	8	0.06	37.11
5.0	3	2.0	12	0.06	51.78

Reproduced with permission from M.M. Alavi Nikje and A. Bagheri Garmarudi, *Polymer–Plastics Technology and Engineering*, 2006, **45**, 10, 1101. ©2006, Taylor & Francis [1]

Table 4.8: Density of samples containing Rezol® IL800 in foam formulations.

Virgin polyol (g)	Recycled polyol (g)	Rezol® (g)	MDI (g)	Density (kg/m^3)
6.5	3	0.5	10	39.93
6.0	3	1.0	10.5	37.12
5.5	3	1.5	11	35.13
5	3	2	11.5	Failed

Reproduced with permission from M.M. Alavi Nikje and A. Bagheri Garmarudi, *Polymer–Plastics Technology and Engineering*, 2006, **45**, 10, 1101. ©2006, Taylor & Francis [1]

polyurethane (PU) scraps in the presence of glycols and catalysts, and carrying out transesterification reactions to obtain low-Mw products.

Reports have demonstrated that microwave (MW) irradiation can accelerate chemical reactions and increase the rate factor to >1,000. MW heating arises from the ability of some liquids and solids to transform absorbed electromagnetic energy into heat. In the MW method, energy can be applied directly to the sample rather than conductively *via* a vessel. Heating can be started or stopped instantly, or the power level can be adjusted. If the sample is mixed efficiently during MW irradiation, bulk heating occurs. The effect of MW on chemical reactions is usually

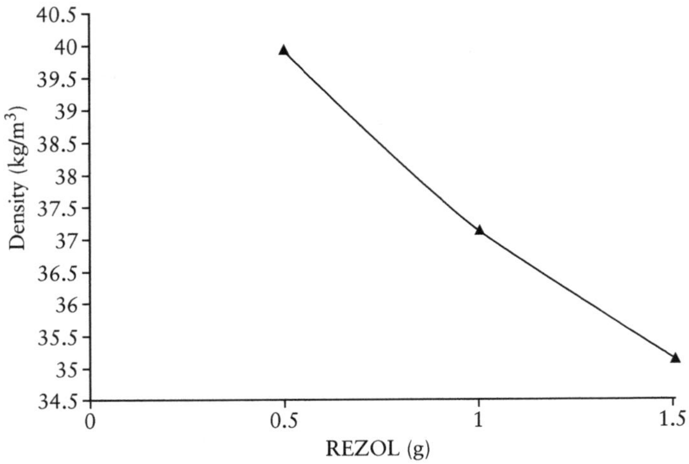

Figure 4.4: Density comparison of different Rezol® IL800 blended polyol foams.
Reproduced with permission from M.M. Alavi Nikje and A. Bagheri Garmarudi,
Polymer–Plastics Technology and Engineering, 2006, **45**, 10, 1101.©2006, Taylor & Francis [1].

investigated by comparing the time required to obtain a given yield of the final products with results from conventional heating methods.

In this work, glycolysis of rigid PUF was carried out using MW irradiation at atmospheric pressure. Effects of different metal hydroxide and acetate catalysts, as well as various MW power values, were investigated [2].

4.2.2 Experimental section

Samples of rigid PUF were formulated using virgin (not recycled) polyol (DALTOFOAM® TA 14,066) at 100 parts by weight (pbw) and MDI-based prepolymer (SUPRASEC® 5005) at 133 pbw.

Fourier-Transform infrared (FTIR) spectra were run using a Tensor 27 spectrometer (Bruker). MW irradiations were carried out using a domestic MW oven (Samsung). The prefix 'virgin' was employed to differentiate the first-time used polyol from the recycled polyol. The virgin polyol was provided as a system containing all the additives required for foaming. Rigid PUF wastes were glycolysed using DEG as the solvent and different catalysts containing NaOH, KOH, sodium acetate (NaAc) and zinc acetate [$Zn(Ac)_2$]. The solvent:foam ratio was 2:1. In a typical process, rigid PUF wastes (3 g) were added to a mixture of DEG (6 g) and catalyst (0.03 g). The reaction mixture was placed in a 100-ml Pyrex™ beaker and treated for a specified time in the MW oven at atmospheric pressure.

In all experiments, foams were smeared in the destroying solvent before the reaction was started. The reaction was done at MW power values of 180, 300, 450, 600, and 900 W to study the effect of different power values on foam dissolution rate. In all reactions, complete disappearance of PUF particles in the mixture was set as the reaction completion time. Hence, all reactions were allowed to proceed until complete dissolution of the foam. Then, the reaction container was removed from the MW oven, solid contaminants and unreacted particles were filtered, and the prepared liquid product was characterised. FTIR spectroscopy studies and reactivity factors were compared with a rigid PUF manufactured by a virgin polyol.

4.2.3 Results and discussion

The glycolysis of rigid PUF led to a hydroxyl-containing product that was liquid at RT. The chemistry of the reaction involved transurethanisation of urethane bonds by a low-Mw glycol Figure 4.5. As shown in the figure, there were two routes for incision of urethane bonds. The first mechanism produced a hydroxyl-containing urethane oligomer and was followed until liberation of the main polyol. The second mechanism was C-N bond cleavage that led to an unstable carbonate and aromatic amine.

Obtained results representing the kinetics of the catalytic reaction was different to a non-catalytic reaction. The transesterification reaction was hard to carry out without a catalyst, even at high power values of MW. Conversely, comparison of foam dissolution times in the presence of different catalysts revealed the weak catalytic effect of $Zn(Ac)_2$ in comparison with other catalysts. NaOH and KOH are convenient catalysts in transesterification reactions that can accelerate glycolysis and reduce the dissolution time (Figure 4.6).

FTIR spectra demonstrated that incision of PU chains was not completed even when using a convenient catalyst, sufficient MW power, or long-time reactions. That is, there were certain amounts of PU oligomers in the reaction mixture due to urethane and biuret bonds (1,760–1,749 cm^{-1}) in recovered products. Absorbance intensity of the remaining functional groups in recycled products decreased if the irradiation was continued after perfect dissolution of the foam Figure 4.7.

Another important observation in these reactions was the compatibility of the recycled polyol with the virgin polyol which caused it to be usable in a similar formulation for manufacture of a new rigid PUF. Accordingly, the recovered polyol was mixed directly in various ratios in combination with a virgin polyol consisting of activated polyether polyols, aliphatic amines catalysts, surfactant and water. The virgin recycled polyol blend was then mixed with isocyanate in a free rise condition (cup test) for manufacture of rigid PUF (Table 4.9).

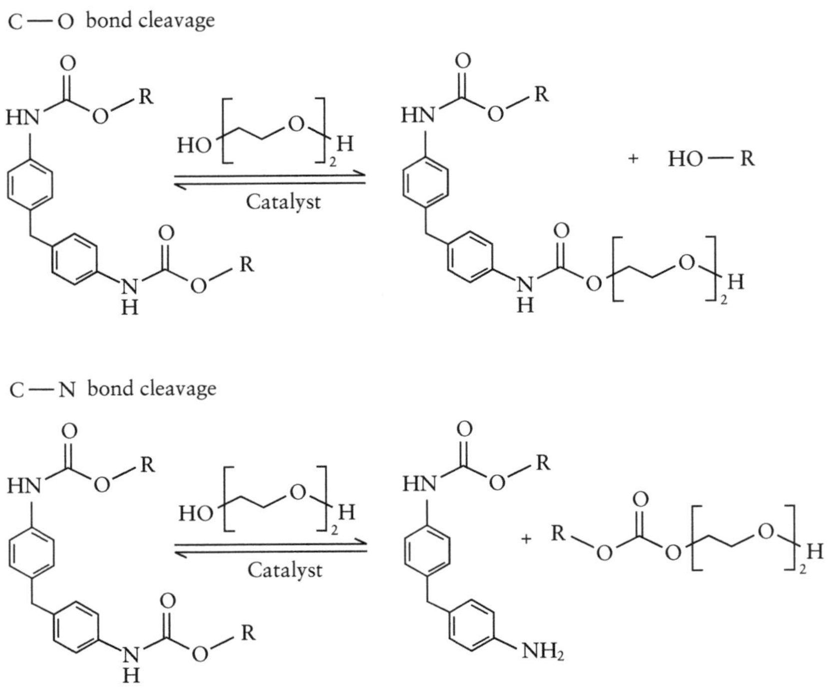

Figure 4.5: Transurethanisation mechanism of urethane bonds by low-Mw glycols. Reproduced with permission from M.M. Alavi Nikje and M. Nikrah, *Journal of Macromolecular Science, Part A: Pure and Applied Chemistry*, 2007, **44**, 6, 613.©2007, Taylor & Francis [2].

Reactivity factors (cream, string, end of rise, and tack-free times) and densities of obtained foams were measured. All data obtained from these tests are detailed in Tables 4.10 and 4.11, respectively.

As shown in Table 4.10, reactivity parameters were reduced by increasing the content of recycled polyol in the polyol blend. Conversely, the foaming process of blends containing high recyclate contents was more exothermic and the heat release during foaming was enhanced by increasing the recyclate content in the polyol blend. The density of the synthesised foam also increased with increasing recyclate content in the virgin recycled polyol blend.

4.3 Chemical recycling of semi-rigid polyurethane foams using an environmentally friendly and green method

4.3.1 Introduction and background

Recycling of polymeric products is investigated to find a suitable method for waste reduction, protect the environment and prevent waste landfilling. Traditional and

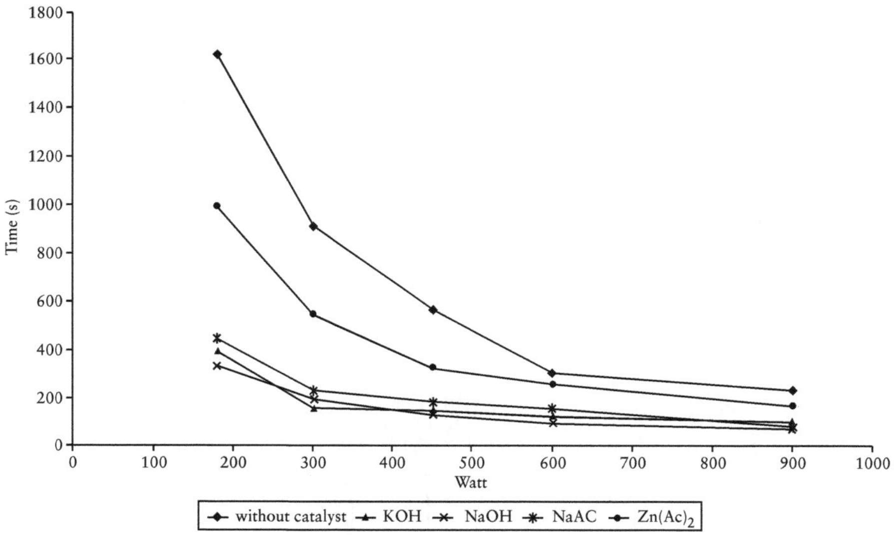

Figure 4.6: Comparison of the various catalysts and values of MW power on foam dissolution time. Reproduced with permission from M.M. Alavi Nikje and M. Nikrah, *Journal of Macromolecular Science, Part A: Pure and Applied Chemistry*, 2007, **44**, 6, 613. ©2007, Taylor & Francis [2].

uncontrolled waste-destroying procedures are not acceptable due to combination of the recycling with pollution of air and water. Hence, chemical-recycling processes have been proposed in the last three decades. Among chemical- recycling processes, glycolysis is popular because it is simple and economical. In this research, we used a sorbitol/DEG/water ternary system as a new and environmentally friendly solvent–reactant mixture to study the efficiency of polyol recovery [3, 4].

4.3.2 Experimental section

Polyol (Daltorim® EK 20,350) and MDI (SUPRASEC® 2082) were used for an integral skin PUF formulation at 100 and 41 portions. Sorbitol solution (70%) was also used.

Scraps of integral skin PUF (15 g) segmented into small sizes, base catalyst (1% *w/w*) and solvent were poured into a three-necked flask equipped with a mechanical stirrer and a removable reflux condenser (Table 4.12). Heating was carried out using an oil bath by setting the oil at 195 ± 5 °C, and the mixture was stirred at 1,000 rpm. The transesterification reaction was carried out at atmospheric pressure and the reaction extended to complete dissolution of foams. At the end of the reaction, the mixture was transferred to a decanted funnel and left to cool at RT. Then, the product was separated into two phases. The upper layer was the recycled polyol and identified by spectroscopy methods.

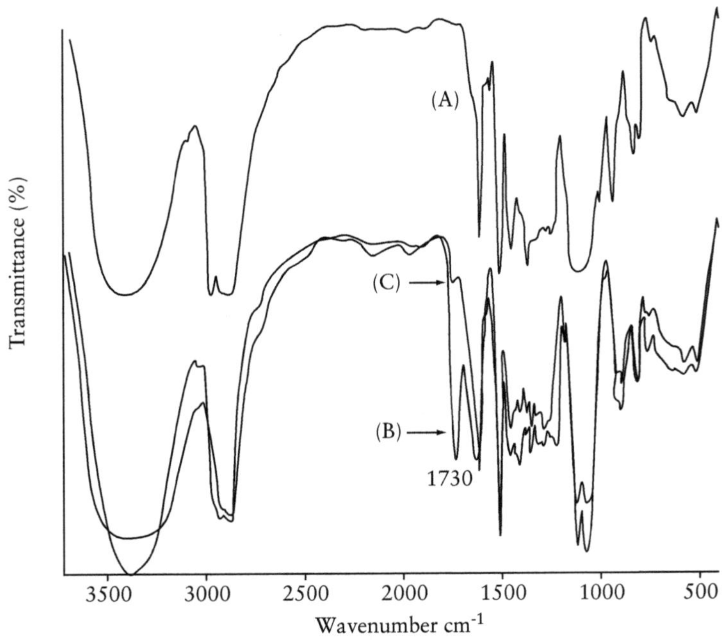

Figure 4.7: FTIR spectra of (A) DALTOFOAM® TA 14,066; (B) recovered polyol sampled immediately after perfect foam dissolution; and (C) 60 s after perfect foam dissolution using NaOH as the catalyst at 900 watt MW power. Reproduced with permission from M.M. Alavi Nikje and M. Nikrah, *Journal of Macromolecular Science, Part A: Pure and Applied Chemistry*, 2007, **44**, 6, 613. ©2007, Taylor & Francis [2].

Table 4.9: Formulations of rigid PUF.

Sample	Virgin polyola	Recycled polyol	MDIb
SPUF1	100	0	100
SPUF2	90	10	100
SPUF3	80	20	100
SPUF4	70	30	100
SPUF5	60	40	100
SPUF6	50	50	100
SPUF7	40	60	100

a DALTOFOAM® TA 14,066.

b SUPRASEC® 5005.

Reproduced with permission from M.M. Alavi Nikje and A. Bagheri Garmarudi, *Polymer–Plastics Technology and Engineering*, 2006, **45**, 10, 1101. ©2006, Taylor & Francis [1]

Table 4.10: Reactivity factors of PUF samples during formation.

Sample	Cream time (s)	String time (s)	End of rise time (s)	Tack-free time (s)
SPUF1	4	45	82	83
SPUF2	4	34	54	55
SPUF3	6	31	51	51
SPUF4	4	25	30	31
SPUF5	4	20	36	36
SPUF6	4	–	25	25
SPUF7	5	–	20	20

Reproduced with permission from M.M. Alavi Nikje and M. Nikrah, *Journal of Macromolecular Science, Part A: Pure and Applied Chemistry*, 2007, **44**, 6, 613. ©2007, Taylor & Francis [2]

Table 4.11: Density of free-rise PUF samples.

Sample	Density (g/m3)
SPUF1	0.0386
SPUF2	0.0448
SPUF3	0.0466
SPUF4	0.0614
SPUF5	0.0712
SPUF6	0.1012
SPUF7	0.1154

Reproduced with permission from M.M. Alavi Nikje and M. Nikrah, *Journal of Macromolecular Science, Part A: Pure and Applied Chemistry*, 2007, **44**, 6, 613. ©2007, Taylor & Francis [2]

Table 4.12: Effects of the composition of the solvent system on split-phase ratios.

Sample	DEG (% w/w)	Sorbitol (% w/w)	Water (% w/w)	Reaction time (min)	Upper-phase (%)	Lower-phase (%)
GSF1	95	3.5	1.5	8	42	58
GSF2	90	7.0	3.0	9	50	50
GSF3	85	10.5	4.5	9	50	50
GSF4	80	14.0	6.0	10	54	46
GSF5	75	17.5	7.5	10	58	42

4.3.3 Results and discussion

The main aim of this research was to use sorbitol as a portion of a destroying solvent system in the recycling process. Hence, to investigate the effects of sorbitol on glycolysis, we examined the dependence of reaction times on sorbitol content using various basic catalysts. We found out that with increasing sorbitol content, the reaction time increased. This phenomenon was possibly due to the increasing viscosity of the reaction media and strong intra- and intermolecular hydrogen-bond formation between sorbitol molecules and/or other hydroxyl content compounds. When the viscosity increased, chain motion for attachment of nucleophiles became restricted and the reaction rate dropped significantly. In the meantime, to investigate the role of the base catalyst in foam dissolution, the first reaction was carried out in the uncatalysed condition, and data suggested non-completion of the reaction (even at longer reaction times). This observation revealed the role of basic catalysts in the glycolysis reaction. That is, for complete and adequate foam dissolution, the reaction should be catalysed using suitable bases. Hence, various catalysts were examined in combination with the solvent system to reduce the dissolution time. To overcome the increasing viscosity of the media, the solvent system was held as DEG (95% wt), sorbitol (3.5% wt) and water (1.5% wt) in combination with different catalysts in 1 pbw ratios. Results suggested complete dissolution of foam at adequate reaction times in the presence of the bases studied, but NaOH decreased the dissolution time significantly. All data are shown in Table 4.13.

Table 4.13: Dissolution time of ISPUF by different catalysts.

Catalyst	Dissolution time (min)
NaOH	8
KOH	9
NaAc	11
Potassium carbonate	9

ISPUF: Integral skin polyurethane foam

Another goal of this research was reduction of consumed energy by avoiding mass losses. Several tests were done to find the best reaction temperature using NaOH as a catalyst. The best reaction temperature was selected to be 200–205 °C. Results for adjustment of the reaction parameters are shown in Table 4.14.

Figures 4.8 and 4.9 show the reaction mechanism of the breakdown of PU bonds by DEG and sorbitol anions as the effective nucleophiles, respectively. As shown in the figures, the first step was the conversion of DEG and sorbitol molecules to the corresponding anions, and further attachment of the anions to PU functional groups.

Table 4.14: Dissolution studies of ISPUF at different temperatures.

Solution No.	PUF(g)	Solvent (g)	Temperature (°C)	Time (min)	Result
1	15	15	150 ± 5	20	Foam not dissolved
2	15	15	180 ± 5	20	Foam dissolved in 14 min
3	15	15	190 ± 5	20	Foam dissolved in 8 min
4	15	15	205 ± 5	20	Excellent dissolution with solvent evaporation

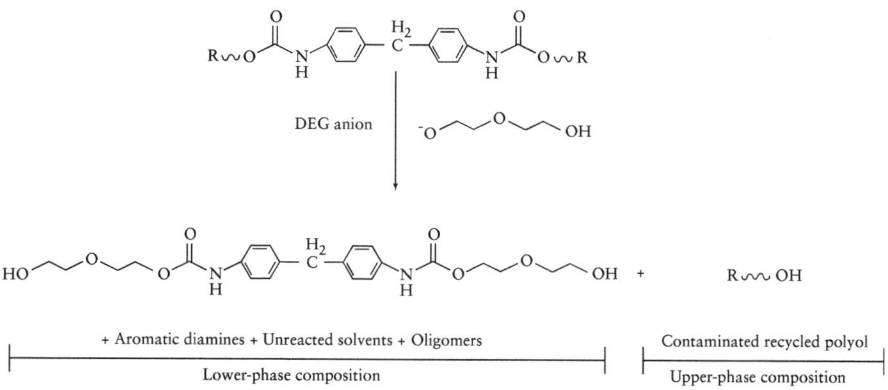

Figure 4.8: Cleavage of PU bonds by DEG anions.

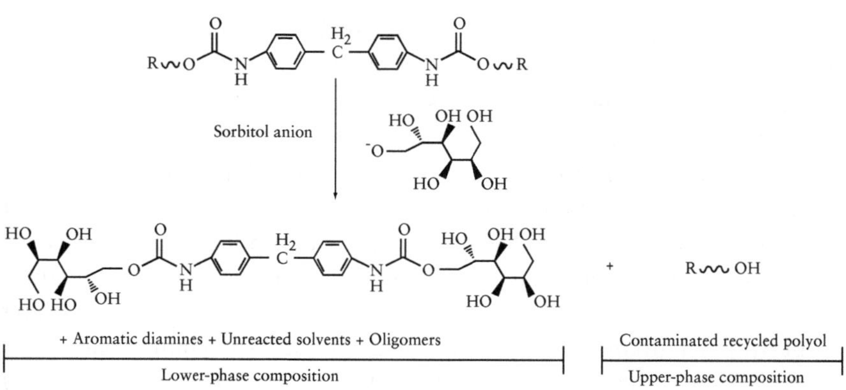

Figure 4.9: Cleavage of PU bonds using sorbitol anions.

Results of breaking of PU bonds were liberation of the polyol as a valued material in combination with other functionalised chemicals.

^{1}H-nuclear magnetic resonance (NMR) spectra of virgin and recycled polyols are shown in Figure 4.10. The peak in the region 1.1 parts per million (ppm) is related to methyl-equivalent hydrogen atoms, and the peaks in 3.0–4.0 ppm regions correspond to hydrogen atoms in the β-position to oxygen atoms. Comparison of the ^{1}H-NMR spectra of the virgin polyol and recycled product in the upper-phase suggested that the chemical structures of the compared compounds were very similar, except for the peaks at 6.6–7.1 ppm and 6.9–7.2 ppm (Figure 4.11b), which corresponded to the aromatic-ring hydrogens that remained as aromatic diamines and contaminated the upper-phase.

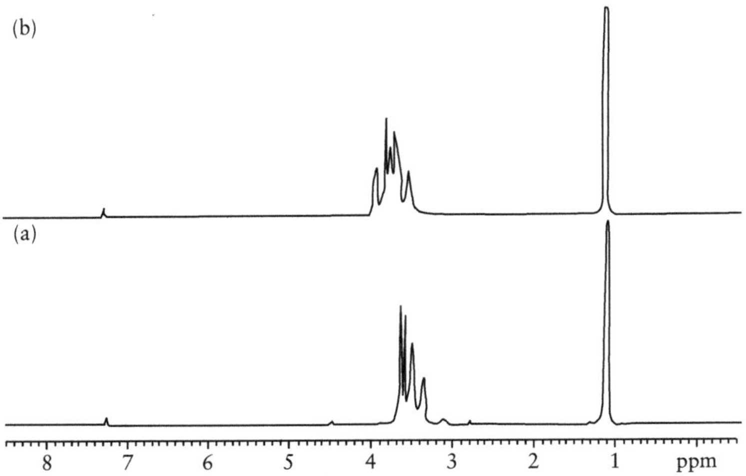

Figure 4.10: ^{1}H-NMR spectra of virgin (a) and recycled (b) polyols.

^{13}C-NMR spectra of recycled polyols at various conditions were quite similar to the virgin polyol except for five new peaks at 40, 114.5, 130.5, 129 and 146 ppm caused by aromatic diamine compounds slightly dissolved in the upper-phase (Figures 4.12–4.14).

FTIR spectroscopy is also used to confirm successful recycling processes and contamination of the recyclate by aromatic diamines. Figure 4.15 shows the FTIR spectra of a contaminated recyclate. In the FTIR spectrum, the absorption band at 1,115 cm^{-1} corresponded to an aliphatic ether group of the polyether polyol. Bending vibrations of methylene groups appeared in the polyol chain at 1,374 and 1,456 cm^{-1}, stretching vibrations of hydroxyl groups were found at 3,482 cm^{-1} and stretching vibrations of CH bonds in aliphatic carbons were found at 2,970–2,868 cm^{-1}. The expected results from the FTIR spectrum were similar to those of the NMR spectrum, and some

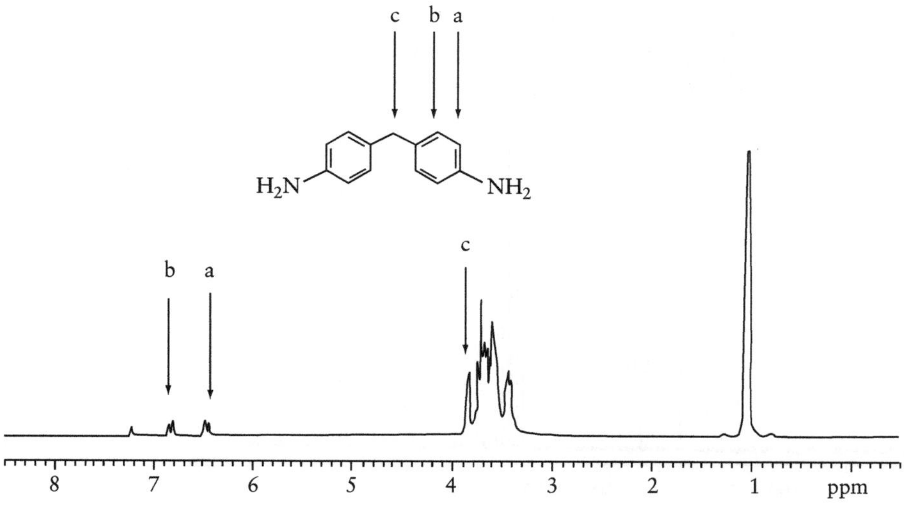

Figure 4.11: ^{1}H-NMR spectrum of contaminated polyols (aromatic diamines).

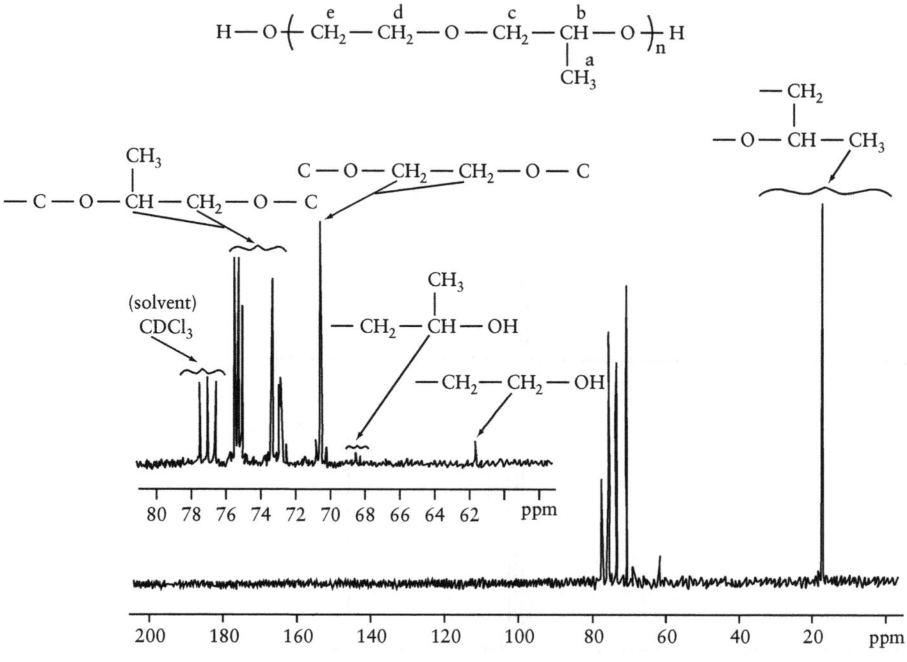

Figure 4.12: ^{13}C-NMR spectrum of integral skin virgin polyol.

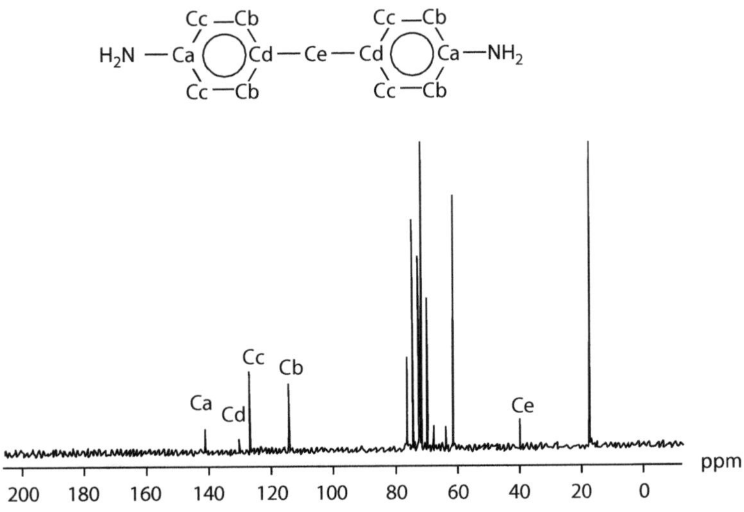

Figure 4.13: ^{13}C-NMR spectrum of contaminated integral skin recycled polyols.

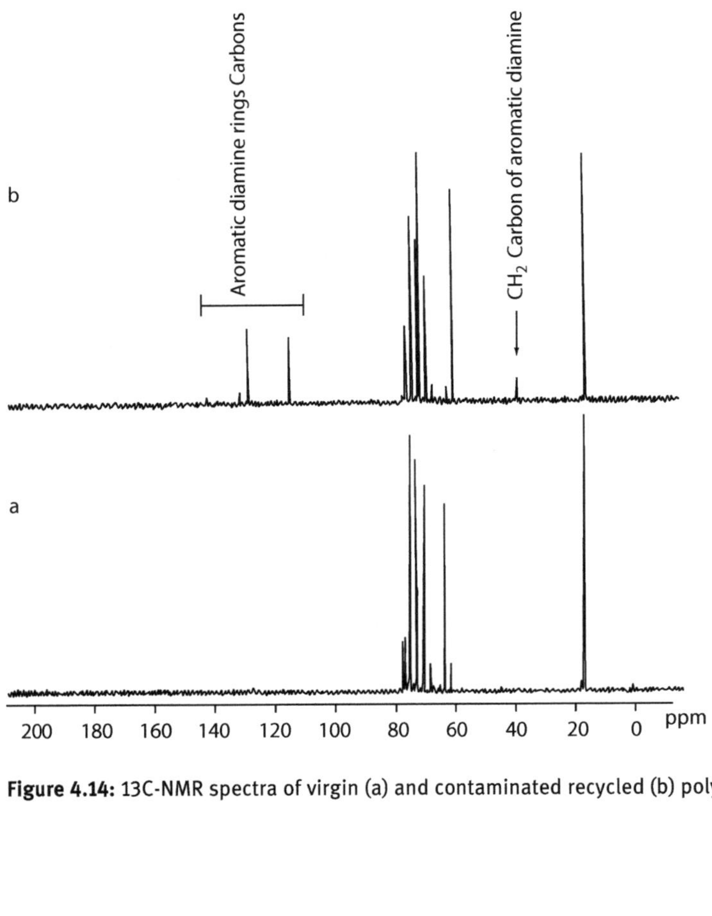

Figure 4.14: 13C-NMR spectra of virgin (a) and contaminated recycled (b) polyols.

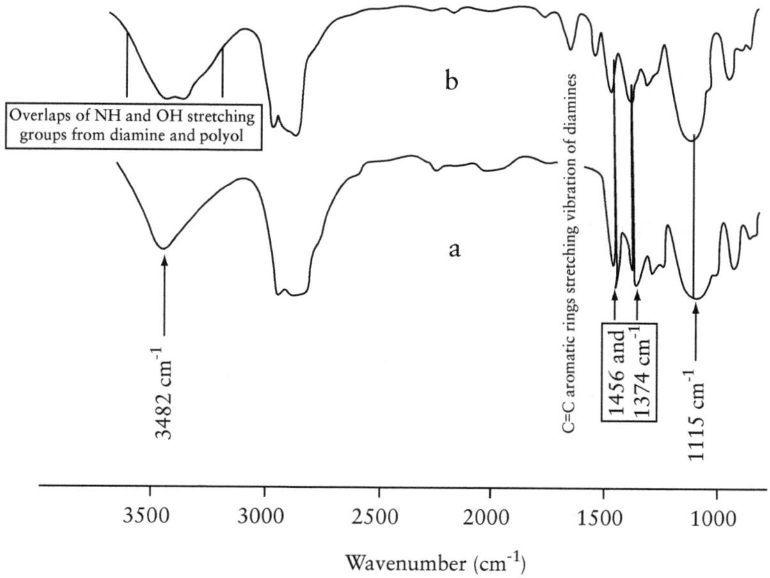

Figure 4.15: FTIR spectra of virgin (a) and amine-contaminated recycled (b) polyols.

additional signals appeared due to aromatic diamine contaminants in combination with the respective signals of the virgin polyol.

There are two mechanistic alternatives for formation of aromatic diamines. The first mechanism is the reaction of water with NCO groups and formation of the unstable intermediate carbamic acid, which results in elimination of carbon dioxide and formation of aromatic diamines. The second mechanism is cleavage of C-N bonds, which results in an unstable carbonate and aromatic amine. The proposed mechanisms are shown in Figure 4.16.

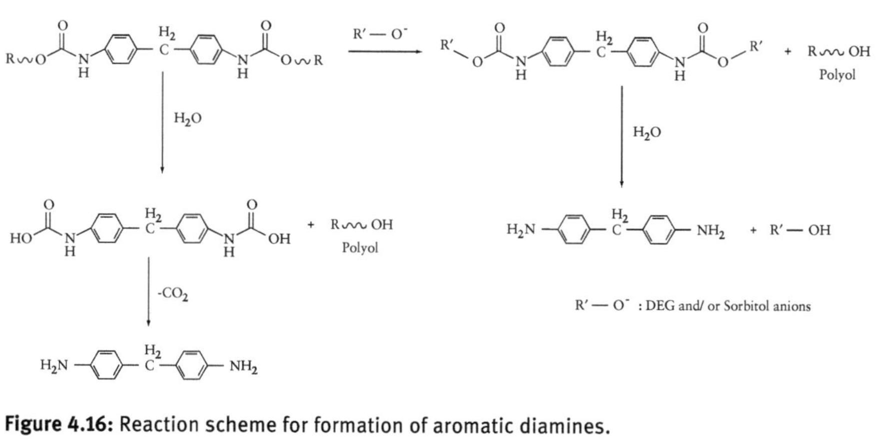

Figure 4.16: Reaction scheme for formation of aromatic diamines.

References

1. M.M. Alavi Nikje and A. Bagheri Garmarudi, *Polymer– Plastics Technology and Engineering*, 2006, **45**, 10, 1101.
2. M.M. Alavi Nikje and M. Nikrah, *Journal of Macromolecular Science, Part A: Pure and Applied Chemistry*, 2007, **44**, 6, 613.
3. M.M. Alavi Nikje and K.M. Tavassoli, *Current Chemistry Letters*, 2012, **1**, 4, 175.
4. M.M. Alavi Nikje in *Automotive Polyurethanes: For Performance and Profit*, Amsterdam, The Netherlands, 2014.

Abbreviations

1,4-BD	1,4-Butanediol
ASTM	American Society for Testing and Materials
$CDCl_3$	Deuterated chloroform
KOH	Potassium hydroxide
CFC	Chlorofluorocarbon(s)
CO_2	Carbon dioxide
DEA	Diethanolamine
DEG	Diethylene glycol
DETA	Diethylenetriamine
EA	Ethanolamine
EG	Ethylene glycol
EMEA	Europe, the Middle East and Africa
FGI	Functional group interconversion
FTIR	Fourier-Transform infrared
GPC	Gel permeation chromatography
$H_{12}MDI$	4,4′-Diisocyanato dicyclohexylmethane
HDI	1,6-Hexamethylene diisocyanate
HMTA	Hexamethylenetetramine
HPLC	High-performance liquid chromatography
IPDI	Isophorone diisocyanate
IR	Infrared
ISPUF	Integral skin polyurethane foam
KOH	Potassium hydroxide
MDA	4,4′-Methylenedianiline
MDI	4,4′-Diphenylmethane diisocyanate
MW	Microwave
Mw	Molecular weight
MwD	Molecular weight distribution
NaAc	Sodium acetate
NaOH	Sodium hydroxide
NMR	Nuclear magnetic resonance
OH#	Hydroxyl number(s)
pbw	Parts by weight
PCL	Polycaprolactone(s)
PEG	Polyethylene glycol
PER	Pentaerythritol
PO	Propylene oxide
ppm	Parts per million
PU	Polyurethane(s)
PUF	Polyurethane foam(s)
RI	Refractive index
RIM	Reaction injection moulding
RT	Room temperature
T_d	Decomposition temperature of 10% of sample mass
TDA	Toluene diamine
TDI	Toluene diisocyanate

https://doi.org/10.1515/9783110641592-005

TGA	Thermogravimetric analysis
THF	Tetrahydrofuran
TPU	Thermoplastic polyurethane
UV	Ultraviolet
$Zn(Ac)_2$	Zinc acetate

Index

https://doi.org/10.1515/9783110641592-006

Bei Fragen zur Produktsicherheit wenden Sie sich bitte an:
If you have any questions regarding product safety,
please contact:

Walter de Gruyter GmbH
Genthiner Straße 13
10785 Berlin
productsafety@degruyterbrill.com